4/96

ILLUSTRATED PHYSIOLOGY

ISBN 0 443 00705 5

First Edition -	-	-	-	-	1963
Revised Reprint -	-	-	-	-	1965
Italian translation -	-	-	-	-	1966
Reprinted -	-	-	-	-	1967
Second Edition -	-	-	-	-	1970
Reprinted -	-	-	-	-	1972

Printed in the British Commonwealth

ILLUSTRATED PHYSIOLOGY

BY

ANN B. MᶜNAUGHT Gilmour

M.B., Cʜ.B., Pʜ.D.

Lecturer in Physiology,
The University of Glasgow.

AND

ROBIN CALLANDER

F.F. Pʜ., M.M.A.A.

Supervisor of Unit for Medical
Illustration and Photography,
The University of Glasgow.

SECOND EDITION
(REPRINT)

The Williams & Wilkins Company

Baltimore

1972

THIS BOOK
IS
AFFECTIONATELY
DEDICATED

To my husband, James A. Gilmour

Ann B. McKnight

To my wife, Elizabeth

Robin Callander

PREFACE
FIRST EDITION

This book grew originally from the need to provide visual aids for the large number of students in this department who come to the study of human physiology with no background of mammalian anatomy and often without any conventional training in either the biological or the physical and chemical sciences. Such students include post-graduates studying for the Diploma or Degree in Education; under-graduates working for Degrees in Science; laboratory technicians taking courses for the Ordinary and Higher National Certificates in Biology; and the increasing number of medical auxiliaries — physio-therapists, occupational therapists, radiographers, cardiographers, dietitians, almoners and social workers — many of whom are required to study to quite advanced levels at least regional parts of the subject.

Many medical, dental and pharmacy students, as well as nurses in training, have been good enough to indicate that they too find diagrams which summarize the salient points of each topic valuable aids to learning or revision. It is our hope that some of these groups may find our book helpful.

Each page is complete in itself and has been designed to oppose its neighbour. It is hoped that this will facilitate the choice of those pages thought suitable for any one course while making it easy to omit those which are too detailed for immediate consideration.

A book of this sort is largely derivative and it is impossible to acknowledge our wider debt. We wish to record our gratitude, however, to Professor R.C. Garry for his generous permission to borrow freely from the large collection of teaching diagrams built up over the years by himself and his staff; to Dr H.S.D. Garven and Dr G. Leaf for permission to use some of their own teaching material; and to Messrs Ciba Pharmaceutical

v

Products Inc. from whose fine book of Medical Illustrations by Dr Frank H. Netter the diagram on the Cranial Nerves has been modified.

We are indebted to the following colleagues and friends who read parts of the original draft and offered helpful criticism:- Dr H.S.D. Garven, Dr J.S. Gillespie, Mr J.A. Gilmour, Dr M. Holmes, Dr B.R. Mackenna, Mr T. McClurg Anderson, Dr I.A. Boyd, Dr R.Y. Thomson, Dr J.B. deV. Weir.

We should like to express our gratitude to Mr Charles Macmillan and Mr James Parker of Messrs. E. & S. Livingstone Ltd. for their unfailing courtesy and encouragement, and to Mrs Elizabeth Callander for help in preparing the index.

ANN B. McNAUGHT

ROBIN CALLANDER.

*Institute of Physiology,
The University of Glasgow.*

January, 1963.

SECOND EDITION

We have been much encouraged by the favourable reception accorded "Illustrated Physiology" and are again indebted to reviewers and colleagues for suggested improvements. In the 1965 revision we recorded our thanks to Professor E.A. Dawes, Reckitt Professor of Biochemistry in the University of Hull, and to Drs. T.D.M. Roberts and N.C. Spurway of the Institute of Physiology, University of Glasgow. On this occasion we wish to add the names of Dr. G.S. Dawes, Director of the Nuffield Institute for Medical Research, the University of Oxford, and Drs. J. Womersley and G. Stenhouse of this department.

A.B. McN.
R.C.

1970

CONTENTS

CHAPTER 1

INTRODUCTION:

The TISSUES

The AMOEBA

All living things are made of PROTOPLASM. Protoplasm exists in MICROSCOPIC UNITS called CELLS. The SIMPLEST living creatures consist of ONE CELL. The AMOEBA (which lives in pond water) exemplifies the BASIC STRUCTURE of all animal cells and shows the PHENOMENA which distinguish living from non-living things.

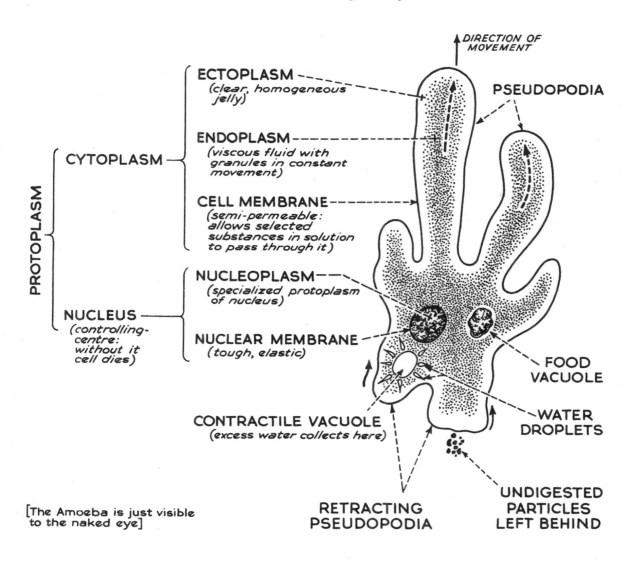

DIRECTION OF MOVEMENT

ECTOPLASM ---
(clear, homogeneous jelly)

PSEUDOPODIA

ENDOPLASM ---
(viscous fluid with granules in constant movement)

CYTOPLASM

CELL MEMBRANE ---
(semi-permeable: allows selected substances in solution to pass through it)

PROTOPLASM

NUCLEOPLASM ---
(specialized protoplasm of nucleus)

NUCLEUS
(controlling-centre: without it cell dies)

NUCLEAR MEMBRANE ---
(tough, elastic)

FOOD VACUOLE

WATER DROPLETS

CONTRACTILE VACUOLE
(excess water collects here)

RETRACTING PSEUDOPODIA

UNDIGESTED PARTICLES LEFT BEHIND

[The Amoeba is just visible to the naked eye]

3

The PHENOMENA which characterize all living things ---- are shown by the AMOEBA

① **ORGANIZATION**
Autoregulation —
inherent ability to control
all life processes.

② **IRRITABILITY**
Ability to respond to
stimuli (from changes
in the environment).

③ **CONTRACTILITY**
Ability to move.

④ **NUTRITION**
Ability to ingest,
digest, absorb
and assimilate
          food.

⑤ **METABOLISM and GROWTH**
Ability to liberate potential
energy of food and to convert
it into mechanical work
(*e.g. movement*) and to
rebuild simple absorbed
units into the complex
protoplasm of the living cell.

⑥ **RESPIRATION**
Ability to take in oxygen
for oxidation of food
with release of energy;
and to eliminate the
resulting carbon
dioxide.

⑦ **EXCRETION**
Ability to eliminate
waste products of
metabolism.

⑧ **REPRODUCTION**
Ability to reproduce the
          species.

PSEUDOPODIA FORMATION

Small elevation arises on surface — Cytoplasm streams forward — forming long pseudopodium

②,③

Free floating — Contact made — Crawling along surface
Attracted by favourable stimuli.
Repelled by unfavourable stimuli.

FOOD VACUOLE FORMATION

INGESTION: *Amoeba flows round and engulfs food particle.*

④

SECRETION of Enzymes (*chemical agents*) which DIGEST (*break down*) complex foods. — ABSORPTION & UTILIZATION of simple units by the living cell. — EXPULSION of undigested particles.

RESPIRATION and EXCRETION

⑥,⑦

Oxygen (*in solution*) — Waste Products (*in solution*)
Carbon Dioxide (*in solution*) — Water periodically expelled at surface

REPRODUCTION   Asexual in the amoeba — by simple fission.

⑧

The PARAMECIUM

The Paramecium (another ONE-CELLED fresh water creature) shows:—

MODIFICATION and localization
of STRUCTURE ----- for ---- SPECIALIZATION and
localization of
CERTAIN FUNCTIONS

DIRECTION OF
MOVEMENT

CONTRACTILE
VACUOLE

CYTOPLASM

CELL
MEMBRANE

NUCLEUS

CONTRACTILE
VACUOLE

[The Paramecium
is just visible to
the naked eye]

CONSTANT
SHAPE
(oval, flat,
slightly twisted)

GROOVE

CILIA
(hair-like
processes)

GULLET
(one definite
place for
intake of food)

FOOD
enclosed in
VACUOLES
in which it is
digested

UNDIGESTED
REMAINS

(a) MOVEMENT
Body as a whole
is propelled through
water in a spiral
course due to the
lashing action of the
cilia.

(b) NUTRITION
Water currents
created.
Food particles
carried into gullet.

Always follow the
same pathway within
the cytoplasm.

Always discarded
at the same outlet
on cell surface.

*Many other one-celled animals show
elaborate organization and specialization.*

5

The CELL

— STRUCTURAL *and* FUNCTIONAL *unit of the* MANY-CELLED ANIMAL.
Higher animals, including MAN, are made up of millions of living cells which vary widely in structure and function but have certain features in common.

A GENERALIZED ANIMAL CELL *(of secretory type).* [The finer details are seen only in the high magnification afforded by ELECTRON MICROSCOPE.]

CYTOPLASM

CELL MEMBRANE
(semi-permeable: controls intake and output of soluble substances)

HYALOPLASM
(clear, viscous matrix: particles in constant movement)

ORGANELLES
(integral parts of the cytoplasm concerned in essential processes of cell's activity)

(a) GOLGI BODY
(concerned with concentrating certain secretions)

(b) CENTROSOME
(with two centrioles– initiates Cell Division)

(c) MITOCHONDRIA
(energy generating centres of cell)

(d) ENDOPLASMIC RETICULUM *(ergasto-plasm)* with RIBOSOMES *(sites of protein synthesis)*

(e) LYSOSOMES
[The limiting membrane ruptures if cell (or part of it) is damaged. Enzymes escape which help to digest the injured tissue]

INERT INCLUSIONS

SECRETION granules

NUCLEUS

NUCLEAR MEMBRANE
(with pores)

NUCLEOPLASM
(specialized protoplasm of nucleus)

CHROMATIN masses
(rich in deoxyribonucleic acid [DNA]) give rise, in cell division, to a constant number of CHROMOSOMES — on these are the GENES, the carriers of inheritable characteristics.

NUCLEOLUS
(rich in ribonucleic acid [RNA])

FAT globules
PROTEIN masses
GLYCOGEN granules
} ---- *'Fuel' materials being stored.*

SECRETION products
WASTE materials
(by-products of cell's activities)
} -- *Awaiting expulsion from cell.*

6

CELL DIVISION (*MITOSIS*)

All cells arise from the division of pre-existing cells. In MITOSIS there is an exact QUALITATIVE division of the NUCLEUS and a less exact QUANTITATIVE division of the CYTOPLASM.

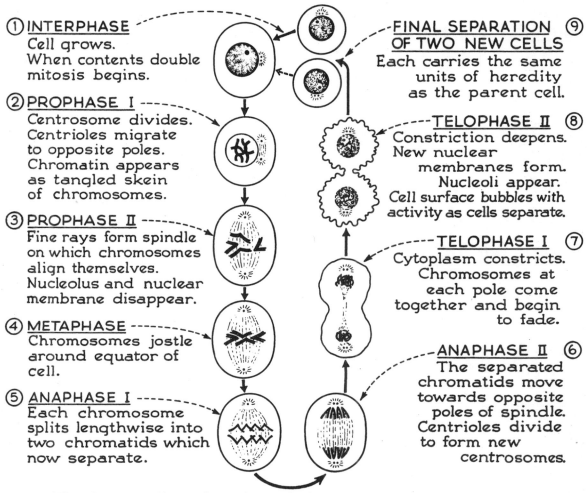

①INTERPHASE
Cell grows.
When contents double mitosis begins.

②PROPHASE I
Centrosome divides.
Centrioles migrate to opposite poles.
Chromatin appears as tangled skein of chromosomes.

③PROPHASE II
Fine rays form spindle on which chromosomes align themselves.
Nucleolus and nuclear membrane disappear.

④METAPHASE
Chromosomes jostle around equator of cell.

⑤ANAPHASE I
Each chromosome splits lengthwise into two chromatids which now separate.

FINAL SEPARATION ⑨
OF TWO NEW CELLS
Each carries the same units of heredity as the parent cell.

TELOPHASE II ⑧
Constriction deepens.
New nuclear membranes form.
Nucleoli appear.
Cell surface bubbles with activity as cells separate.

TELOPHASE I ⑦
Cytoplasm constricts.
Chromosomes at each pole come together and begin to fade.

ANAPHASE II ⑥
The separated chromatids move towards opposite poles of spindle.
Centrioles divide to form new centrosomes.

The longitudinal halving of chromosomes and genes ensures that each new cell receives the same hereditary factors as the original cell.

The number of chromosomes is constant for any one species.

The cells of the human body (somatic cells) carry 23 pairs — i.e. 46 chromosomes.

[For clarity only 4 chromosomes (2 pairs) are shown in these diagrams]

DIFFERENTIATION of ANIMAL CELLS

SPECIALIZATION distinguishes multicellular creatures from more primitive forms of life.

ONE-CELLED ANIMALS — Capable of INDEPENDENT existence — *Undifferentiated* — Show all activities or ······PHENOMENA of LIFE

MANY-CELLED ANIMALS — Cells CO-OPERATE for well-being of whole body.

Differentiated —— Groups of cells undergo adaptations and sacrifice some powers to fit them for special duties.

All cells retain powers of
ORGANIZATION
IRRITABILITY
NUTRITION
METABOLISM
RESPIRATION
EXCRETION

MODIFICATION of STRUCTURE ---*for efficient*--- SPECIALIZATION of FUNCTION ---*with*--- LOSS or REDUCTION of VERSATILITY

E.g.

SECRETORY CELL ········· Highly developed powers of SECRETION e.g. enzymes for chemical breakdown of foodstuffs.

Diminished powers of CONTRACTION and REPRODUCTION

Cytoplasm

Nucleus displaced to base by formed and stored secretion

Nucleus

FAT CELL ---------------- STORAGE of FAT

Loss of powers of CONTRACTION and SECRETION

Cytoplasm

Cytoplasm displaced by stored fat

Nucleus

MUSCLE CELL ------------- Highly developed powers of CONTRACTILITY

Diminished powers of SECRETION and REPRODUCTION

Cytoplasm   Nucleus

Elongated cell body

NERVE CELL ------------- Highly developed powers of IRRITABILITY

Cytoplasm

(response to stimuli and transmission of impulses over long distances)

Loss of powers of REPRODUCTION

i.e. if nerve cell is destroyed no regeneration is possible.

Cytoplasm drawn out into long branching processes

Nucleus x500

8

ORGANIZATION OF TISSUES

Different cell types are not mixed haphazardly in the body.
Cells which are alike are arranged together to form TISSUES.
There are *four* main types of tissue:– 1. EPITHELIA or LINING,
2. CONNECTIVE or SUPPORTING, 3. MUSCULAR, 4. NERVOUS.

EPITHELIA

STRUCTURAL MODIFICATIONS	SITE	SPECIALIZED FUNCTIONS

Sheets of cells with minimum intercellular substance

Line all internal and external surfaces of body

A. SIMPLE — Single layer

(a) SQUAMOUS

Lining of blood vessels, heart and cavities of body.

Reduces friction between surfaces.

(b) CUBICAL

SIMPLE — Small ducts e.g. of salivary glands.

Protects underlying tissues: non-secretory.

SECRETORY – SEROUS — Many glands e.g. serous secreting part of mixed salivary gland.

Forms watery secretion containing e.g. digestive enzymes.

MUCOUS — Mucous glands e.g. in mixed salivary gland.

Forms viscous lubricant, protective mucus.

(c) COLUMNAR

Simple — Large ducts of kidney.

Protects and lines.

Mucous — Surface lining of stomach.

Secretes carpet of mucus to protect stomach from its own acid and digestive enzymes.

Goblet Cells — E.g. in lining of intestine.

Form a viscous, lubricating, protective mucus.

Striated or Brush Bordered — Small intestine.

Absorbs foodstuffs.

Ciliated *(Motile hair-like processes at free surface)* — Uterine tube.

Forms currents — wafts ovum towards womb.

x500

9

EPITHELIA

B. PSEUDOSTRATIFIED COLUMNAR CILIATED *(with Goblet cells)* ------- Respiratory passages.

(a) Protective.

(b) Cilia form currents to move mucus-trapped particles towards mouth.

C. COMPOUND - Many layers of cells

TRUE STRATIFIED

(a) *TRANSITIONAL* --------------------- Urinary passages.

(a) Extensible.

(b) Protective — prevents penetration of urine into underlying tissues.

(b) *STRATIFIED SQUAMOUS* ------------ Surfaces subjected to great wear and tear.

Most resistant type of epithelium.

(i) *Uncornified.*

Internal surfaces e.g. mouth and gullet.

Protects against friction.

(ii) *Cornified* ----------- External skin surfaces.

Protects against wear and tear, evaporation and extremes of temperature.

Cornified dead cell layers

Clear layer

-anular layer

Prickle cell layers

Germinal layer (for cell replacement)

[Dead cell layers vary in thickness — e.g. thickest on soles of feet and palms of hands.]

×500

10

CONNECTIVE TISSUES

STRUCTURAL MODIFICATIONS
Cells plus large amount of intercellular matrix and extracellular elements.

SPECIALIZED FUNCTIONS
Form framework, connecting, supporting and packing tissues of the body.

1. CELLS (floating free) in a FLUID MATRIX *(without fibrils until it clots).*
   <u>BLOOD</u> Lymph

   Red Blood Corpuscle *(no nucleus)* ⎯ ⎯ ⎯ ⎯ ⎯ ⎯ ⎯ ⎯ ⎯ ⎯ ⎯ Oxygen transport to tissues.

   White Blood Cells

   Granular .... | *Neutrophil* ⎯ ⎯ ⎯ ⎯ ⎯ ⎯ ⎯ ⎯ ⎯ ⎯ Defence against bacteria.
   Eosinophil ⎯ ⎯ ⎯ ⎯ ⎯ ⎯ ⎯ ⎯ ⎯ Rôle in allergy.
   Basophil ⎯ ⎯ ⎯ ⎯ ⎯ ⎯ Heparin formation.

   Monocyte ⎯ ⎯ ⎯ ⎯ ⎯ ⎯ Defence against bacteria.

   Non-granular {
   Lymphocytes } Immunity reactions (Antibody formation).

   ×1000

   Platelets *(no nuclei)* ⎯ ⎯ ⎯ ⎯ ⎯ ⎯ ⎯ ⎯ ⎯ ⎯ Rôle in blood clotting.

2. CELLS in SEMI-SOLID JELLY-LIKE MATRIX with fine fibrils.
   <u>MESENCHYME</u>

   Mesenchyme cells

   Cytoplasm drawn out to touch neighbouring cells

   Earliest type found in embryo.

   From it all other connective tissues differentiate.

3. CELLS in SEMI-SOLID MATRIX with fine network of extracellular RETICULAR FIBRES.
   <u>RETICULAR</u>

   Reticular fibres ⎯ ⎯ ⎯ Form 3-dimensional 'net'. Framework of e.g. spleen, lymph nodes, bone marrow.

   Reticular cells ⎯ ⎯ Potential phagocytes - capable of ingesting particles.

   ×500

CONNECTIVE TISSUES

4. CELLS in SEMI-SOLID MATRIX with thicker collagenous
and elastic fibres.

(a) LOOSE FIBROUS

Fibroblasts (cells actively forming fibres)

Collagen fibres

Elastic fibres

"Packing" tissue between organs: sheaths of muscles, nerves and blood vessels.

(b) DENSE FIBROUS

Fibrocytes (resting cells)

Collagen fibres

Elastic fibres

Strong, inelastic yet pliable— e.g. ligaments; capsules of joints; heart valves.

(c) ELASTIC

Elastic fibres (or laminae)

Strong, extensible and flexible— e.g. in walls of blood vessels and air passages.

(d) ADIPOSE

Fat filled cells

Supporting collagen fibres

Protective "cushion" for organs. Insulating layer in skin. Storage of fat reserves.

(e) TENDON

Tendon cells

Dense fibrous tissue bundles

Tough, inelastic cords of dense fibrous tissue which attach muscles to bones.

x 500

12

CONNECTIVE TISSUES

5. CELLS in SOLID ELASTIC MATRIX with fibres.

(a) HYALINE CARTILAGE

Cartilage cells

Matrix apparently homogeneous

Firm yet resilient
e.g. in air passages; ends of bones at joints.
Transition stage in bone
                      formation.

(b) WHITE FIBRO-CARTILAGE

Cartilage cells

Collagen fibres in matrix

Tough. Resistant to stretching.
Acts as shock absorber between vertebrae.

(c) ELASTIC or YELLOW FIBRO-CARTILAGE

Cartilage cells

Elastic fibres in matrix

More flexible, resilient —
        e.g. in larynx and ear.

×500

6. CELLS in SOLID RIGID MATRIX impregnated with Calcium and
                              Magnesium Salts.
BONE

Lamellae

Bone cell

Forms rigid framework
        of body.

×500

Haversian canal containing blood vessels.
×100

13

MUSCULAR TISSUES

All have ELONGATED cells with special development of CONTRACTILITY.

1. SMOOTH, UNSTRIPED, VISCERAL or INVOLUNTARY muscle

Nucleus

Least specialized. Shows slow rhythmical contraction and relaxation.
Not under voluntary control.
Found in walls of viscera and blood vessels.

2. CARDIAC or HEART muscle

Nucleus

Cross-striation

More highly specialized.
Rapid rhythmical contraction (and relaxation) spreads through whole muscle mass.
Not under voluntary control.
Found only in heart wall.

Cells adhere, end to end, at intercalated discs to form long 'fibres' which branch and connect with adjacent 'fibres'.

(Note: There is no protoplasmic continuity between cells.)

3. STRIATED, STRIPED, SKELETAL or VOLUNTARY muscle

Nucleus

Cross-striated fibrils

x500

Most highly specialized.
Very rapid, powerful contractions of individual fibres.
Under voluntary control.
Found in e.g. muscles of trunk, limbs, head.

Thick covering membrane (sarcolemma)

Very long, large multi-nucleated units. No branching.

NERVOUS TISSUES

Nervous tissue is divided into:-

(a) <u>NEURONES</u> or *True* nerve cells specialized in
IRRITABILITY
CONDUCTION
INTEGRATION

(b) <u>ACCESSORY</u> or *Supporting* cells
Not RECEPTIVE
Not CONDUCTING

<u>MOTOR</u> — pass messages from brain and spinal cord to effector organs (muscles and glands)

<u>ASSOCIATION</u> — relay messages between neurones.

<u>SENSORY</u> — receive and pass messages from environment to brain and spinal cord.

<u>NEUROGLIA</u> in Central Nervous System (*Brain and Spinal Cord*)

<u>SHEATH</u> (*Schwann*) cells on peripheral nerve fibres, i.e. outside C.N.S.

<u>SATELLITE</u> cells in ganglia of peripheral nervous system.

- -

<u>TYPICAL NERVE CELL</u> (*Motor*)

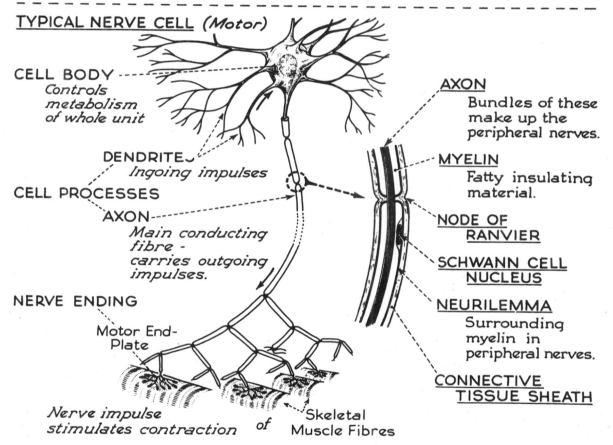

<u>CELL BODY</u>
Controls metabolism of whole unit

<u>DENDRITE</u>
Ingoing impulses

<u>CELL PROCESSES</u>

<u>AXON</u>
Main conducting fibre - carries outgoing impulses.

<u>NERVE ENDING</u>

Motor End-Plate

<u>AXON</u>
Bundles of these make up the peripheral nerves.

<u>MYELIN</u>
Fatty insulating material.

<u>NODE OF RANVIER</u>

<u>SCHWANN CELL NUCLEUS</u>

<u>NEURILEMMA</u>
Surrounding myelin in peripheral nerves.

<u>CONNECTIVE TISSUE SHEATH</u>

Nerve impulse stimulates contraction of Skeletal Muscle Fibres

NERVOUS TISSUES

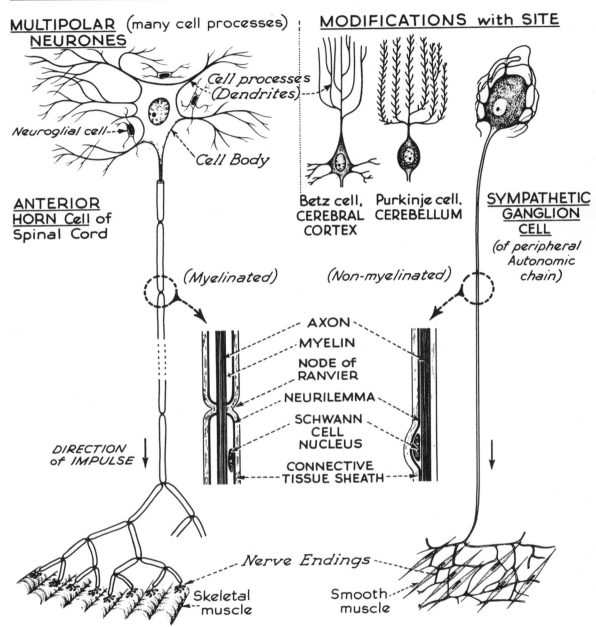

MULTIPOLAR NEURONES (many cell processes)

Neuroglial cell

Cell processes (Dendrites)

Cell Body

ANTERIOR HORN Cell of Spinal Cord

(Myelinated)

DIRECTION of IMPULSE

MODIFICATIONS with SITE

Betz cell, CEREBRAL CORTEX

Purkinje cell, CEREBELLUM

(Non-myelinated)

SYMPATHETIC GANGLION CELL

(of peripheral Autonomic chain)

AXON
MYELIN
NODE of RANVIER
NEURILEMMA
SCHWANN CELL NUCLEUS
CONNECTIVE TISSUE SHEATH

Nerve Endings

Skeletal muscle

Smooth muscle

Most multipolar neurones are MOTOR (efferent) or ASSOCIATION in function.

16

NERVOUS TISSUES

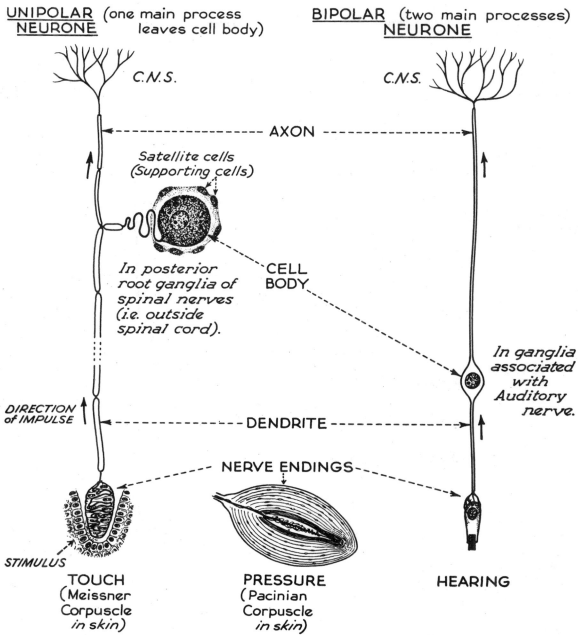

UNIPOLAR (one main process
NEURONE   leaves cell body)

BIPOLAR (two main processes)
NEURONE

C.N.S.

C.N.S.

AXON

Satellite cells
(Supporting cells)

*In posterior
root ganglia of
spinal nerves
(i.e. outside
spinal cord).*

CELL
BODY

*In ganglia
associated
with
Auditory
nerve.*

DIRECTION
of IMPULSE

DENDRITE

NERVE ENDINGS

STIMULUS

TOUCH
(Meissner
Corpuscle
in skin)

PRESSURE
(Pacinian
Corpuscle
in skin)

HEARING

All unipolar and bipolar neurones are SENSORY (afferent) in function.

CELL DIVISION *(MEIOSIS)*

New individuals develop after fusion of 2 specialized cells—the GAMETES. During their formation the OVUM *(female)* and the SPERMATOZOON *(male)* undergo two special cell divisions to reduce the chromosome content of each to the haploid number of 23. In fusion, the mingling of male and female chromosomes restores the normal diploid number—46.

MATURATION of the MALE GAMETES *(only one pair of chromosomes illustrated)*

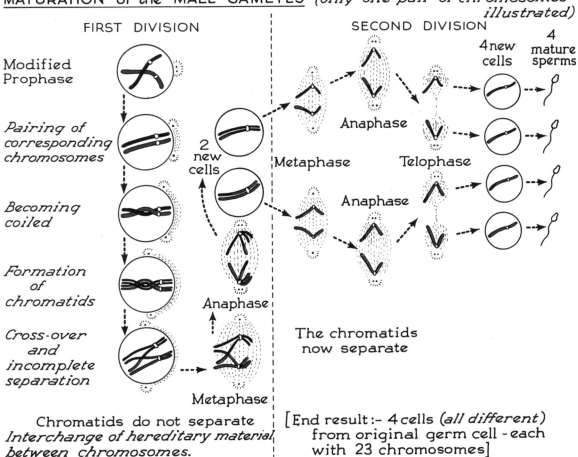

FIRST DIVISION

Modified Prophase

Pairing of corresponding chromosomes

2 new cells

Metaphase

SECOND DIVISION

4 new cells

4 mature sperms

Anaphase

Telophase

Becoming coiled

Formation of chromatids

Anaphase

Anaphase

The chromatids now separate

Cross-over and incomplete separation

Metaphase

Chromatids do not separate *Interchange of hereditary material between chromosomes.*

[End result:- 4 cells *(all different)* from original germ cell - each with 23 chromosomes]

In the **FEMALE** , three of the 'cells' are small polar bodies which are discarded and disintegrate.

ONE single mature OVUM receives most of the cytoplasm.

DEVELOPMENT of the INDIVIDUAL

All tissues of the human body are derived from the single cell—
the FERTILIZED OVUM.

FERTILIZATION
Fusion of ovum and spermatozoon

2 cells   4 cells   8 cells   Ball of 16 cells

CLEAVAGE
Repeated mitotic divisions: each cell
receives equal number of *Maternal* and
Paternal chromosomes.

Blastocyst–hollow sphere

BLASTULATION and IMPLANTATION in womb   DIFFERENTIATION of tissues

EMBRYONIC GERM DISC

{ *ECTODERM* gives rise to ----Epithelia of *External* surfaces.
----Nervous tissues.

{ *MESODERM* gives rise to ----Muscular tissues.
----Connective tissues.
----Urogenital system.
----Lining of body cavities and blood vessels.

{ *ENDODERM* gives rise to ----Epithelia of most *Internal* surfaces.
----Some glands (e.g. liver, pancreas, thyroid).

"MASTER" TISSUES

It is an aid to understanding the general Design of the body to think of the systems as grouped into "MASTER" Tissues and "VEGETATIVE" Systems.

The "MASTER" Tissues are those which specialize in receiving stimuli from the environment and reacting to them —
NERVOUS TISSUE and SKELETAL MUSCLE.

NERVOUS SYSTEM

EXTERNAL
ENVIRONMENT
ACTING ON
ORGANISM

PERIPHERAL RECEPTORS
AND
INGOING NERVES

CENTRAL
CONTROLLING
CENTRES

IN

BRAIN

AND

SPINAL
CORD

SOUND,
LIGHT → TELERECEPTORS
are stimulated by
stimuli from a
distance

DIRECT
CONTACT → CONTACT
RECEPTORS
at surface of
body are
stimulated

INTERNAL
ENVIRONMENT

MOVEMENT
in MUSCLES → PROPRIOCEPTORS
are stimulated

*
MOVEMENT
in VISCERA → INTEROCEPTORS
are stimulated

VISCERAL
MUSCLE

[*AUTONOMIC
NERVOUS
SYSTEM]

Reaction of
Organism to ← Outgoing Nerves
Environment   to
Effectors — MUSCLES

[*Note that a special part of the Nervous System — the AUTONOMIC NERVOUS SYSTEM — regulates the activities of the "Vegetative" Systems.]

20

"VEGETATIVE" SYSTEMS

The VEGETATIVE Systems are those which serve the basic utilities of life. They deal with the source of energy — FOOD- breaking it down and transporting it around the body to make it available to the cells for Growth and Repair, for Reproduction and for Movement, etc., and get rid of Waste. Much of their activity goes on below the level of consciousness.

DIGESTIVE SYSTEM

FOOD
the source of energy

Digestion

SMALL INTESTINE

Digestion, Absorption to Blood Stream.

LARGE INTESTINE

Reabsorption of water and salt

FAECES
Waste residue

STOMACH

LIVER

TRANSPORT SYSTEM
LUNGS
HEART

ENDOCRINE GLANDS
A REGULATING SYSTEM

KIDNEY

KIDNEY

BLADDER

RESPIRATORY SYSTEM

O$_2$ for combustion of food and release of energy

Expiration of CO$_2$ + H$_2$O

EXCRETORY SYSTEM

KIDNEYS
Filtration of blood. Reabsorption of essential substances. Maintenance of alkalinity and composition of body fluids.

SKIN   Protection. Control of heat loss. Excretion of some waste products in sweat.

URINE
Waste products of metabolism

The activities of these systems are regulated partly through a section of the MASTER tissues - the AUTONOMIC NERVOUS SYSTEM. Its complementary divisions - PARASYMPATHETIC and SYMPATHETIC - carry opposite or antagonistic messages to viscera. These impulses are rapidly adjusted and balanced by the BRAIN CENTRES to integrate the actions of the vegetative systems to the constantly changing needs of the whole body.

The BODY SYSTEMS

The tissues are arranged to form ORGANS.
Organs are grouped into SYSTEMS.

The ESSENTIAL LIFE ----are delegated to------- SEPARATE
PROCESSES SYSTEMS

1. IRRITABILITY and CONTROL (response to stimuli: and integration)

2. METABOLISM and GROWTH (energy release for work and growth)

3. RESPIRATION (oxygen intake for release of energy; CO_2 loss)

4. TRANSPORT of materials (waste, food, respiratory gases)

5. NUTRITION (source of energy ingested, digested and absorbed)

6. EXCRETION (waste products of metabolism eliminated)

7. REPRODUCTION (propagation of species)

8. CONTRACTILITY (movement)

CENTRAL NERVOUS

ENDOCRINE

RESPIRATORY

CARDIO-VASCULAR

DIGESTIVE

EXCRETORY

REPRODUCTIVE

LOCOMOTOR (skeletal tissues: muscles)

The systems do not work independently. The body works as a whole. Health and well-being depend on the co-ordinated effort of every part.

22

CHAPTER 2

NUTRITION and METABOLISM

The SOURCES, RELEASE and
USES of ENERGY

BASIC CONSTITUENTS of PROTOPLASM

Protoplasm is made up of certain
ELEMENTS —— present mainly in —— **CHEMICAL COMBINATION**

Chemical Symbol ⌄ | ⌄ | *e.g. in a 70 kg man average amounts (grams)* ⌄

H – HYDROGEN	6,580	A large amount of the H and O is present as WATER	C, H & O combine chemically to form CARBOHYDRATES and FATS *(chief sources of ENERGY in living protoplasm)*	C, H, O and N combine to form PROTEINS *(main BUILDING constituents of all protoplasm)*	*These make up most of the Body Weight*
O – OXYGEN	43,550				
C – **CARBON**	12,590				
N – NITROGEN	1,815				

Ca – CALCIUM	1,700	Important constituents of blood and of hard tissues – e.g. bones and teeth.	*These eight make up much of the remaining Body Weight*
P – PHOSPHORUS	680		
Cl – CHLORINE	115	Important constituents of body fluids.	
Na – SODIUM	70		
K – POTASSIUM	70	Important constituent of all cells.	
S – SULPHUR	100		
Mg – MAGNESIUM	42	Important for activity of Brain, Nerves and Muscles.	
Fe – IRON	7	Important component of red blood cells.	

<u>Trace elements</u> *make up the last few grams or so. These include:- Manganese, copper, iodine, zinc, cobalt, molybdenum, nickel, aluminium, chromium, titanium, silicon, rubidium, lithium, arsenic, fluorine, bromine, selenium, boron, barium and strontium.*

NOTE:- *Very few of these inorganic substances are found free in protoplasm. Most are in chemical combination. Apart from water, the chief constituents are present as compounds of CARBON, i.e. they are ORGANIC SUBSTANCES.*

CARBOHYDRATES

CARBOHYDRATES are found in Plant and Animal Protoplasm. The simpler carbohydrates are known as SUGARS; the more complex as STARCH. Starch is made up of hundreds of units of sugar tied chemically together. They consist of atoms of C, H and O.

GLUCOSE – $C_6H_{12}O_6$ – is one of the simplest SUGARS — a MONOSACCHARIDE
[Found in plants and animals]

The arrangement of atoms in the molecule may be represented thus:-

usually written →

LACTOSE – $C_{12}H_{22}O_{11}$ – is another relatively simple SUGAR.
[Milk sugar]   Two MONOSACCHARIDE molecules combine to form a –

DISACCHARIDE

with loss of one molecule of WATER (H_2O)

[N.B. Same constituent elements as Glucose but with different spatial arrangement.]

GALACTOSE   GLUCOSE

STARCH – $(C_6H_{10}O_5)_n$ where n>100 – is built up of long chains of GLUCOSE units to form a-
[Form in which PLANTS store sugar]

POLYSACCHARIDE

with loss of one molecule of WATER at each link

These chains may then be arranged in layers:-

chain ——————— link
chain ——————— link
chain

GLYCOGEN – $C_n(H_2O)_m$ – is built up of 12 to 18 GLUCOSE units similarly linked
[Form in which ANIMALS store sugar]   chemically as a POLYSACCHARIDE but with *branching chains*.

FATS

FATS are present in Plant and Animal Tissues. The molecule of fat is made up of smaller molecules of FATTY ACIDS linked chemically with a molecule of GLYCEROL. They consist of atoms of the elements C, H and O.

SIMPLE LIPIDS

GLYCEROL can react with __1__ molecule of a __FATTY ACID__ to produce a __MONOGLYCERIDE__

e.g. *BUTYRIC ACID* — " — MONOBUTYRIN

$$CH_2OH$$
$$CHOH \quad + \quad HO\,OCCH_2CH_2CH_3 \longrightarrow$$
$$CH_2O\,H$$

$$CH_2OH$$
$$CHOH$$
$$CH_2OOCCH_2CH_2CH_3$$

with the loss of one molecule of WATER (H_2O)

__GLYCEROL__ can react with __2__ molecules of __FATTY ACID__ to form a —— __DIGLYCERIDE__

e.g. *PALMITIC ACID* — " — DIPALMITIN

$$CH_2O\,H$$
$$CHOH \quad + \quad HO\,OC\,CH_2CH_2CH_2CH_2CH_2CH_2CH_2CH_2\,CH_2\,CH_2CH_2CH_2CH_2CH_2CH_3$$
$$CH_2O\,H \qquad\qquad HO\,OC\,CH_2CH_2CH_2CH_2CH_2CH_2CH_2CH_2\,CH_2\,CH_2CH_2CH_2CH_2CH_2CH_3$$

—— with the loss of two molecules of H_2O

In its most common form FAT is made up of one molecule of __GLYCEROL__ linked with __3__ molecules of __FATTY ACID__ to form a —— __TRIGLYCERIDE__

e.g. *STEARIC ACID* — " — TRISTEARIN

$$CH_2O\,H \quad HO\,OCCH_2CH_2CH_2CH_2CH_2CH_2CH_2CH_2CH_2CH_2CH_2CH_2CH_2CH_2CH_2CH_2CH_3$$
$$CHO\,H \; + \; HO\,OCCH_2CH_2CH_2CH_2CH_2CH_2CH_2CH_2CH_2CH_2CH_2CH_2CH_2CH_2CH_2CH_2CH_3$$
$$CH_2\,O\,H \quad HO\,OCCH_2CH_2CH_2CH_2CH_2CH_2CH_2CH_2CH_2CH_2CH_2CH_2CH_2CH_2CH_2CH_2CH_3$$

—— with the loss of three molecules of H_2O

There are many other naturally occurring FATTY ACIDS. GLYCEROL can combine with 3 molecules of the __same__ or 3 molecules of *different* FATTY ACIDS to form TRIGLYCERIDES (NEUTRAL FATS).

More complex fat-like substances include

__PHOSPHATIDES__ — *made up of GLYCEROL, FATTY ACIDS, PHOSPHORIC ACID and a NITROGEN base. They are widely distributed in animal tissues.*

__CEREBROSIDES__ — *are even more complex compounds combining SUGAR in their structure.*

COMPOUND LIPIDS

Other fat-like substances (e.g. *Cholesterol, Bile Salts and many Hormones*) are known as __STEROIDS__

They all share this PARENT STRUCTURE

PROTEINS

PROTEIN is the chief organic constituent of all Protoplasm – Plant or Animal. The protein molecule is made up of smaller units called AMINO-ACIDS. These consist of atoms of C, H, O and N with sometimes S and P.

GLYCINE ———————— is the simplest ———————— AMINO-ACID

The arrangement of its atoms may be represented like this:-

[The name 'amino' comes from this NH_2 group] H–N–C–C–OH [The name 'acid' comes from this COOH group]

Both groups are found in all the 20 or so kinds of amino-acid found in nature.

ALANINE ——————— is another simple ——————— AMINO-ACID

H–N–C–C–OH
H (CH₃) O

All amino-acids share the same core- -N-C-C- . Only the group in this position differs in each.

GLYCINE and
ALANINE can combine easily to form a ——————— DIPEPTIDE

(HOH)

H–N–C–C–N–C–C–OH
H (H) O H (CH₃) O

with the loss of one molecule of WATER (H_2O).

AMINO-ACIDS can link in this way many times to form a POLYPEPTIDE

(H₂O) (H₂O) (H₂O) (H₂O)

H–N–C–C–N–C–C–N–C–C–N–C–C–N–C–C–OH
H (R) O H (R) O H (R) O H (R) O H (R) O

each link causing loss of one molecule of WATER (H_2O)

Any of the 20 Amino-acids may be linked like this.   Long chains with 50 to 400 molecules of amino-acids may make up a molecule of PROTEIN.
Note:- The -N-C-C- cores of the amino acids form the backbone of the protein chain.

There are thousands of different kinds of Protein but only about 20 different amino-acids. Differences between proteins depend on the amino-acids present and on their number and arrangement.

The various amino-acids can be present in any proportion, arranged in any order, so that the resulting PROTEIN has its own sequence of groups (R) branching off along the sides — this sequence gives it its own special properties which distinguish it from other proteins. The number of possible variations accounts for the great variety of living species, each with a protoplasm specific to itself.

NUCLEIC ACIDS

NUCLEOPROTEINS *(Protein attached to NUCLEIC ACID)* are amongst the most important constituents of Cells and their Nuclei.

NUCLEIC ACIDS are important in controlling heredity. They have a complex chain-like structure built up from MONONUCLEOTIDES —
e.g. Deoxyadenylic acid.

A
NUCLEOTIDE
contains

$H_2O_3P-O-CH_2$

MOLECULES of:—

Phosphoric —— Sugar —— Purine base
acid   (Deoxyribose-   [Adenine (A)]
            a sugar with   [Guanine (G)]
            5C atoms)   —Pyrimidine base
                            [Cytosine]
                            [Thymine (T)]

These mononucleotides may condense together to form POLYNUCLEOTIDES.

CH_2OH O   Base(A)

PO_4H

CH_2 O   Base(G)

PO_4H

CH_2 O   Base(T)

PO_4H

etc.

Such a chain of nucleotides *(in the DNA molecule in Chromosomes)*, by the specific arrangement of its bases, carries information from one generation of cells to the next.

[The information thus carried in a polynucleotide chain may determine many genetic factors, e.g. whether a child will have brown or blue eyes]

SOURCE of ENERGY : PHOTOSYNTHESIS

The essential life processes or the phenomena which characterize life depend on the use of ENERGY. The SUN is the SOURCE of ENERGY for ALL LIVING THINGS.

Only GREEN PLANTS can TRAP and STORE the sun's energy and build from relatively simple substances the energy-rich and body-building compounds — *CARBOHYDRATES, FATS and PROTEINS* — required by PROTOPLASM. This process is called <u>PHOTOSYNTHESIS</u>.

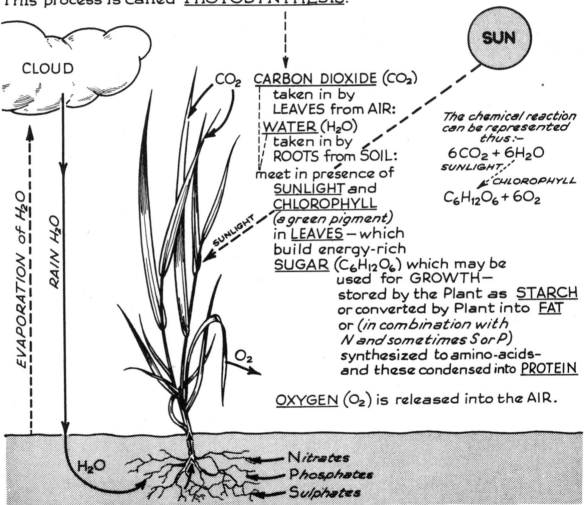

CLOUD

CO_2

CARBON DIOXIDE (CO_2)
  taken in by
LEAVES from AIR:
WATER (H_2O)
  taken in by
ROOTS from SOIL:
meet in presence of
<u>SUNLIGHT</u> and
<u>CHLOROPHYLL</u>
(a green pigment)
in <u>LEAVES</u> — which
build energy-rich
<u>SUGAR</u> ($C_6H_{12}O_6$) which may be
  used for GROWTH—
  stored by the Plant as <u>STARCH</u>
  or converted by Plant into <u>FAT</u>
  or (*in combination with*
  N and sometimes S or P)
  synthesized to amino-acids—
  and these condensed into <u>PROTEIN</u>

<u>OXYGEN</u> (O_2) is released into the AIR.

SUN

The chemical reaction can be represented thus :—

$$6CO_2 + 6H_2O$$
SUNLIGHT
CHLOROPHYLL
$$C_6H_{12}O_6 + 6O_2$$

EVAPORATION of H_2O

RAIN H_2O

SUNLIGHT

O_2

H_2O

Nitrates
Phosphates
Sulphates

When Plants or their products are eaten this stored energy becomes available to animals and man.

CARBON 'CYCLE' in NATURE

Animal bodies are unable to build *Proteins, Carbohydrates* or *Fats* from the raw materials. They must be built up for them by *Plants*.
CARBON is the basic building unit of all these compounds.

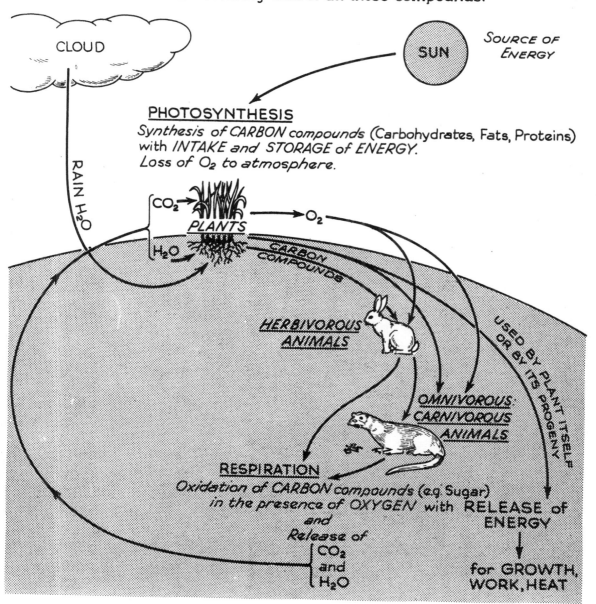

CLOUD

SUN

SOURCE OF ENERGY

PHOTOSYNTHESIS
Synthesis of CARBON compounds (Carbohydrates, Fats, Proteins) with INTAKE and STORAGE of ENERGY.
Loss of O_2 to atmosphere.

RAIN H_2O

CO_2

PLANTS

O_2

H_2O

CARBON COMPOUNDS

HERBIVOROUS ANIMALS

USED BY PLANT ITSELF OR BY ITS PROGENY

OMNIVOROUS CARNIVOROUS ANIMALS

RESPIRATION
Oxidation of CARBON compounds (e.g. Sugar) in the presence of OXYGEN with RELEASE of ENERGY
and
Release of
CO_2
and
H_2O

for GROWTH, WORK, HEAT

NITROGEN 'CYCLE' in NATURE

Although animals are surrounded by NITROGEN in the air they can only use the NITROGEN trapped by Plants.

NITROGEN in the Atmosphere

NUTRITION

Man eats as <u>FOOD</u> the substances which PLANTS
(and, through them, ANIMALS) have made.
These are broken down in man's body
into simpler chemical units which provide

<u>BUILDING</u> and <u>PROTECTIVE MATERIALS</u>

Man requires more or less the same
elements as plants. (Some, such as
the minerals iodine, sodium or iron,
he assimilates in INORGANIC FORM.)

Most must be built up for him
by plants :-
ORGANICALLY COMBINED
Carbon, Nitrogen, and Sulphur, etc.
ESSENTIAL AMINO ACIDS

ESSENTIAL FATTY ACIDS
Certain VITAMINS
These are
<u>USED to</u>
<u>BUILD, MAINTAIN or REPAIR</u>
<u>PROTOPLASM</u>
- -
BODY BUILDING REQUIREMENTS
of the INDIVIDUAL determine
<u>QUALITY</u>
of DIET

ENERGY

Stored originally
by plants

<u>RELEASED</u> in man's cells
by
<u>OXIDATION</u>

When food is 'burned' it gives up
its stored energy

PROTEINS	yield 4·1 kilocalories	*Units of Energy per gram.*
CARBOHYDRATES	yield 4·1 kilocalories	
FATS	yield 9·2 kilocalories	

Most of this appears as <u>HEAT</u>
and is <u>USED for</u>
<u>KEEPING BODY WARM</u>;
some is used for <u>WORK</u> of Cells
- -
ENERGY REQUIREMENTS
of the INDIVIDUAL determine
<u>QUANTITY</u>
of DIET

For a
<u>BALANCED DIET</u>

TOTAL INTAKE
of
Essential Constituents and Energy Units
must balance...
...Amounts *stored* plus amounts *lost* from body plus amounts *used*
as Work or Heat.

ENERGY-GIVING FOODS

All the main foodstuffs yield ENERGY – the energy originally trapped by Plants. CARBOHYDRATES and FATS are the chief energy-giving foods. PROTEINS can give energy but are mainly used for building and repairing protoplasm.

CARBOHYDRATES are the
PRIMARY SOURCE of ENERGY –
More easily and quickly digested
and utilized than Fats

FATS are a
SECONDARY SOURCE of ENERGY –
Ideal energy storage material. Weight
for weight, give twice as many energy
units as Carbohydrates.

PLANT SOURCES:

SUGAR is found in
LEAVES, FRUITS and ROOTS of PLANTS
and in
FOODS made from them by man

JAM, TREACLE
SWEETS, SYRUP

*i.e. especially those products made by plants
for development of next generation*

SUGAR can be stored in plants as
CARBOHYDRATE or converted to FAT and stored as such

STARCH is found in
GRAIN, SEEDS and ROOTS of PLANTS
e.g.
WHEAT

e.g.
POTATOES

OILS in
SEEDS
e.g. OLIVES,
COTTON SEEDS

NUTS
e.g. PEANUTS,
COCONUTS

and in
FOODS made from them by man

FLOUR e.g. BREAD, CAKES.
CEREALS e.g. CORNFLAKES

COOKING FATS

PEANUT BUTTER

ANIMAL SOURCES:

SUGAR (glucose) is found in TISSUES and BLOOD of ANIMALS and in
FOODS made from them by man

e.g. CHOP

e.g. BLACK PUDDING

and in *Products made by Animals for themselves or the next generation.*

MILK SUGAR (lactose) – – – – – – – –

MILK FAT

HONEY (fructose)

MILK – – – – – – – – – – – – → *BUTTER, CHEESE*

SUGAR can be stored in animals as
CARBOHYDRATE or converted to FAT and stored (together with FAT
built from Dietary Fatty Acids and
Glycerol) in FAT DEPOTS of body.

ANIMAL STARCH (glycogen)
is found in *MUSCLES:*
e.g. STEAK

LIVER:

e.g. SUET, LARD, MUTTON FAT, FISH OIL

and in *FOODS made from them by man*

e.g. SAUSAGES

e.g. MARGARINE

BODY-BUILDING FOODS

PROTEIN is the chief body-building food. Because it is the chief constituent of protoplasm, tissues of Plants and Animals are the richest sources.

PLANT SOURCES

Protoplasm of Plant Tissues and — *Stores made by Plants* — or *Foods made*
for the next generation *from them by man.*
in seeds and roots

LEAVES

"2ND CLASS PROTEINS" – a good source of AMINO-ACIDS – but not yielding the full range essential for man's growth

e.g. PEAS,   *POTATOES*
BEANS

e.g. CABBAGE

WHEAT → *e.g.* *FLOUR* → *BREAD*

ANIMAL SOURCES

Protoplasm of Animal Tissues and — *Products made by Animals* — or *Foods made*
to provide for the next *from them by man.*
generation

BEEF

"1ST CLASS PROTEINS" – built up from a range of AMINO-ACIDS similar to those needed for GROWTH or REPAIR of man's own tissues.

MILK

e.g. CHEESE

POULTRY

EGGS

e.g. EGG CUSTARD

FISH

ROE

e.g. MARGARINES
from some fish oils

These foods between them also contain other important body-building elements:- e.g. CALCIUM and PHOSPHORUS for making BONES and TEETH hard. IRON for building HAEMOGLOBIN– the OXYGEN CARRYING substance in RED BLOOD CELLS. IODINE for building the THYROID HORMONE.

PROTECTIVE FOODS — I

Even when provided with BODY-BUILDING and ENERGY-GIVING FOODS, human tissues cannot build or repair their protoplasm or release the energy in the absence of PROTECTIVE substances called <u>VITAMINS</u>. These appear to be important links in METABOLIC processes. Vitamins cannot be substituted for one another in the way that different Carbohydrates, Proteins or Fats can replace others.

ALL are essential for NORMAL GROWTH, HEALTH and WELL-BEING and for general RESISTANCE to INFECTION.

EACH has additional specific protective functions:—

<u>FAT-SOLUBLE VITAMINS</u>	**A** ANTIXEROPH-THALMIC ($C_{20}H_{30}O$)	**D** ANTI-RACHITIC ($C_{28}H_{44}O$)	**E** ANTISTERILITY (Tocopherols $\alpha\beta\gamma$)	**K** ANTI-HAEMORRHAGIC
	Protects SKIN and MUCOUS MEMBRANES (especially front of eye and lining of Digestive and Respiratory tracts). Essential for Regeneration of VISUAL PURPLE.	Important in Ca and P Metabolism. Essential for deposition Ca and P in Bones and Teeth. Promotes absorption of Ca from intestine.	Perhaps important for normal Reproduction in adult. Influences Biochemical Processes in Nuclei of growing cells.	Essential for production of Prothrombin and Factor VII in Liver— important for normal blood clotting.
<u>PLANT SOURCES</u> LEAVES — SEEDS & FRUITS — ROOTS —	Present as precursor CAROTENE. Green, Yellow Vegetables (esp. SPINACH and KALE). Yellow Maize, Peas, Beans. <u>CARROTS</u>.	Vegetables, fruits and cereals } contain negligible amounts.	Green leaves (e.g. LETTUCE) PEAS. Richest source - GERM of various Cereals e.g. <u>WHEAT GERM OIL</u>.	Spinach, Kale, Cabbage, Cauliflower, Cereals, Tomatoes, Carrots, Potatoes.
<u>ANIMAL SOURCES</u> TISSUES — PRODUCTS — for·PROGENY FOODS made — from them	Animal liver breaks down CAROTENE ($C_{40}H_{56} + 2H_2O \rightarrow 2C_{20}H_{30}O$) Stored in LIVER of Animals and <u>FISH</u> Secreted in MILK EGG YOLK BUTTER CREAM	Formed when ultra-violet rays fall on Sterols in Man's Skin. Stored in LIVER of <u>FISH</u> and Animals MILK EGGS BUTTER CREAM CHEESE Vitamin content of dairy produce varies with plant in animal's diet	Small amounts in MEAT and DAIRY PRODUCE	Synthesized by BACTERIA in man's INTESTINE then absorbed in presence of BILE.

COD LIVER OIL

| <u>DIETARY DEFICIENCY</u>

of any or all vitamins retards GROWTH and leads to SEVERE DISEASE and eventually DEATH | <u>XEROPHTHALMIA</u> | <u>RICKETS</u> in CHILD

<u>OSTEOMALACIA</u> in ADULT | Little evidence of DEFICIENCY syndrome in human beings. (In some female animals developing embryos die, and, in male, testes may show degenerative changes) | If DEFICIENT absorption from intestine → upset in mechanism for Blood Clotting |

PROTECTIVE FOODS — II

WATER-SOLUBLE VITAMINS

B Complex

Together form essential parts of tissue ENZYME SYSTEMS - concerned in metabolism and Release of Energy by the cells

Includes
B₁ (Aneurin Thiamine) ANTINEURITIC

B₂ RIBOFLAVIN
NICOTINIC ACID

Each forms part of a COENZYME SYSTEM – concerned with RELEASE of ENERGY from FOOD-STUFFS.

B₆ PYRIDOXINE- *Coenzyme in metabolism of certain amino acids.*

PANTOTHENIC ACID
BIOTIN — *Tissue enzyme in Carbo-hydrate metabolism.*

CHOLINE | *Lipotrophic factors-probably*
INOSITOL | *promote utilization of fats.*
PARA-AMINOBENZOIC ACID – *No known rôle in man.*

FOLIC ACID – An **ANTIANAEMIA** *factor.*

B₁₂ COBALAMINE-ANTIPERNICIOUS ANAEMIA *factor – Involved in maturation of red blood cells*

C ANTISCORBUTIC (C₆H₈O₆)

Probably acts as a Respiratory Catalyst in cells. Essential for formation and maintenance of INTERCELLULAR CEMENT and CONNECTIVE TISSUE. (Especially necessary for healthy state of blood vessels)

PLANT SOURCES

Common natural sources for most members of B Complex:-
PULSES e.g. GREEN PEAS (B₁, B₂)
SEEDS and OUTER COATS OF GRAIN e.g.rice, wheat (B₁, B₂, Nic.A)
NUTS e.g. PEANUTS (B₁)
YEAST and Yeast Extracts (B₁, B₂, Nicotinic Acid)

Green vegetables, e.g. Parsley. Citrus fruits, e.g. oranges, lemons, grapefruits, limes. Tomatoes, Rosehips, Black-currants, Red & Green Peppers. Peas. Turnips, Potatoes.

ANIMAL SOURCES

B₂ and Nicotinic Acid are Synthesized by BACTERIA in the human intestine. Found in LIVER (B₁, B₂, Nicotinic Acid, B₁₂, etc.), BACON and LEAN MEAT, MEAT Extract, MILK (B₂, Nicotinic Acid) EGGS (B₂) CHEESE (B₂)

Stored in body — High concentration in Adrenal Glands. Found in Meat, Liver. Secreted in MILK.

DIETARY DEFICIENCY

Deficiency in Pyridoxine, Pantothenic Acid, Biotin Choline Para-amino benzoic A. not observed in man.

B₁ BERI-BERI
Gross disturbance of function of NERVOUS TISSUE with great MUSCULAR PARALYSIS and WEAKNESS and disturbance in SENSATION. OEDEMA. Gastro-intestinal upsets. HEART FAILURE

B₂ RIBOFLAVINOSIS
Roughening of skin. Cornea becomes cloudy. Cracks and fissures around lips & tongue.

NICOTINIC ACID PELLAGRA
Roughening and reddening of skin. Tongue red and sore. In severe cases-gastro-intestinal upsets and mental derangement

SCURVY
Intercellular cement breaks down. Capillary walls leak → haemorrhages into tissues - e.g. gums swell and bleed easily. Wounds heal slowly.

B₁₂ PERNICIOUS ANAEMIA
Members of B Complex frequently absent together giving grave disturbances of chemical reactions in all tissues.

DIGESTION

The organic substances of man's food are chemically <u>similar</u> to those which his body will form from them. In detail they <u>differ</u>.

Conversion of <u>FOOD SUBSTANCE</u> ——— to ——— <u>BODY SUBSTANCE</u>

Chemical BREAKDOWN
to simpler organic units
in DIGESTIVE TRACT under the action of
ENZYMES

PEPSIN

TRYPSIN
CHYMOTRYPSIN

eg.
PROTEIN: EGG – <u>breakdown</u> → AMINO-ACIDS – – – → Absorption to blood stream.
contains protein–   Synthesis by Body's Cells to
<u>ALBUMIN</u>   e.g. <u>SERUM ALBUMIN</u>

– – – Both Proteins built up – – –
from about 20 Amino-Acids

AMYLASE { Saliva in Mouth
Pancreatic Juice
(Intestinal Juice)

MALTASE

CARBOHYDRATE: POTATO – <u>breakdown</u> → GLUCOSE – – – – – → Absorption: Synthesis by
contains   (MONOSACCHARIDE)   muscle and liver cells to
<u>STARCH</u>   <u>GLYCOGEN</u>

– – Both Polysaccharides built – – –
up from Glucose units

LIPASE

FAT: BUTTER – <u>Partial breakdown</u> → GLYCERIDES – – – → Absorption to lymph or blood
or   stream: Synthesis to
<u>Total breakdown</u> → FATTY ACIDS   <u>BODY FAT</u>
and GLYCEROL

Both Triglycerides built up
from Fatty Acids and Glycerol

DIGESTION *is brought about by* <u>SPECIFIC ENZYMES</u> *themselves made of Protein — each acts as a CATALYST for speeding up one particular chemical breakdown without effect on any others.*

PROTEIN METABOLISM

INGESTION

DIETARY PROTEIN

DIGESTION

broken down in STOMACH by Enzyme: PEPSIN (& HCl) to PEPTONES and PEPTIDES

TRYPSIN in SMALL INTESTINE

AMINO-ACIDS

ABSORPTION

AMINO-ACIDS

Portal Blood

Some deamination by bacteria in LARGE INTESTINE

AMMONIA

NH_3

combined in various ways

EXCRETION
in FAECES

TRANSPORT

SYSTEMIC BLOOD STREAM

rapid exchange without complete breakdown to amino acids

PLASMA PROTEINS

LIVER

Intermediate **METABOLISM**

synthesis → PROTEIN

some stored PROTEIN

AMINO-ACIDS

temporary storage of some

Those which cannot be used by man are broken down under action of Enzymes

e.g HOOC-C-H | R, H N-H, H

Amino group split off (deamination) to form AMMONIA UREA

KETO ACID *oxidized*

$CO_2 + H_2O + ENERGY$

GLUCOSE
GLYCOGEN

FAT → Ketone Bodies

NH_3

AMMONIA
UREA

UREA

NH_3

TISSUES

TISSUE PROTEIN

Other Nitrogenous substances, e.g. HORMONES, PURINES, PYRIMIDINE, CREATINE.

AMINO-ACIDS

temporary storage of some

TISSUE UTILIZATION

Each tissue selects the amino-acids required to build or repair its own characteristic protein

continuous interchange and replacement goes on

AMINO-ACIDS

filtered

reabsorbed

some deamination

AMMONIA

Some also formed in kidney itself

UREA

KIDNEY

NH_3
UREA

derived from DIETARY and TISSUE PROTEIN

EXCRETION
in URINE

PROTEIN METABOLISM *is under control of Hormones – chiefly the* GROWTH HORMONE (SOMATOTROPHIN) *of the* ANTERIOR PITUITARY.

CARBOHYDRATE METABOLISM

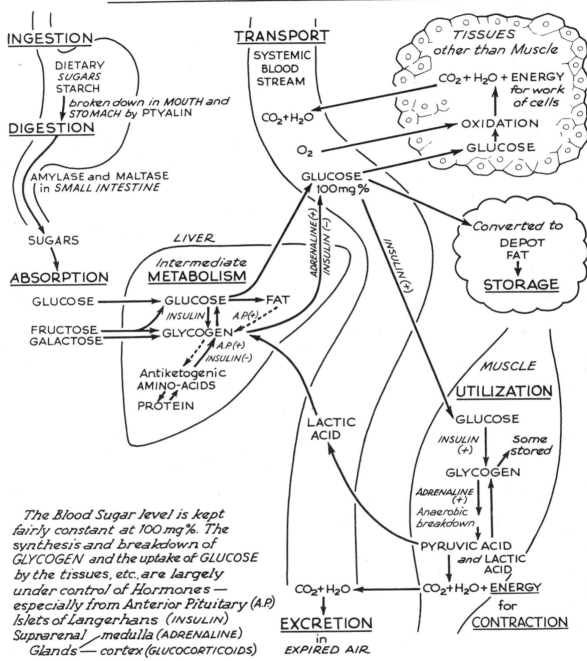

INGESTION

DIETARY *SUGARS STARCH* *broken down in MOUTH and STOMACH by PTYALIN*

DIGESTION

AMYLASE and MALTASE in *SMALL INTESTINE*

SUGARS

ABSORPTION

GLUCOSE

FRUCTOSE
GALACTOSE

LIVER

Intermediate **METABOLISM**

GLUCOSE → FAT
INSULIN   *A.P.(+)*
GLYCOGEN
A.P.(+)
INSULIN(-)

Antiketogenic
AMINO-ACIDS

PROTEIN

TRANSPORT

SYSTEMIC
BLOOD
STREAM

$CO_2 + H_2O$

O_2

GLUCOSE
100mg %

ADRENALINE(+)
INSULIN(-)
INSULIN(+)

LACTIC
ACID

*TISSUES
other than Muscle*

$CO_2 + H_2O$ + ENERGY
*for work
of cells*

OXIDATION

GLUCOSE

Converted to
DEPOT
FAT

STORAGE

MUSCLE

UTILIZATION

GLUCOSE
*INSULIN
(+)*   *Some
stored*

GLYCOGEN
*ADRENALINE
(+)*
*Anaerobic
breakdown*

PYRUVIC ACID
and LACTIC
ACID

$CO_2 + H_2O$ + ENERGY
for
CONTRACTION

$CO_2 + H_2O$

EXCRETION
in
EXPIRED AIR

*The Blood Sugar level is kept
fairly constant at 100 mg %. The
synthesis and breakdown of
GLYCOGEN and the uptake of GLUCOSE
by the tissues, etc., are largely
under control of Hormones —
especially from Anterior Pituitary (A.P.)
Islets of Langerhans (INSULIN)
Suprarenal ⌐medulla (ADRENALINE)
   Glands — cortex (GLUCOCORTICOIDS)*

FAT METABOLISM

INGESTION

DIETARY FAT

Some Neutral (unsplit) fat

DIGESTION

by LYMPHATICS

LIPASE (from PANCREAS)

Partial breakdown to Long chain FATTY ACIDS, MONO- and DI- GLYCERIDES

Complete breakdown to Short chain FATTY ACIDS and GLYCEROL

ABSORPTION

by LYMPHATICS

Portal Blood

TRANSPORT

SYSTEMIC BLOOD STREAM

FAT

INSULIN A.P.

THIAMINE

DEPOT FAT
(especially subcutaneous and retroperitoneal

STORAGE

FAT rebuilt into characteristic human form

TISSUE UTILIZATION

FAT — Some built into PROTOPLASM

Oxidation

O_2 → O_2 + KETONE BODIES

Citric acid cycle

$CO_2 + H_2O$ + ENERGY for work of cells

GLUCOSE

LIVER

Intermediate METABOLISM

FAT → GLYCEROL

GLYCOGEN

INSULIN THIAMINE

FATTY ACIDS

Acetylcoenzyme A

OXIDATION by CITRIC ACID cycle

$CO_2 + H_2O$ + Energy

KETONE BODIES

Ketogenic Amino-Acids

PROTEIN

KETONE BODIES

$CO_2 + H_2O$

KETONE BODIES

KIDNEY

EXCRETION in EXPIRED AIR

EXCRETION in URINE

The Amount of Fat in Blood Stream can vary within wide limits. The storage and oxidation of FAT are largely under the Control of the Endocrine System.

RELEASE of ENERGY

ENERGY is released from FOODSTUFFS by OXIDATION.

ENZYME SYSTEMS exist within cells which can convert (in a series of steps) Fats, Proteins or Carbohydrates into compounds suitable for entering the "ENERGY-PRODUCING CYCLE".

e.g   GLYCOGEN ⟶ GLUCOSE

In the MITOCHONDRIA of the cell is a system of enzymes.

The molecule of "FUEL" is handed on from enzyme to enzyme – each of which changes it a little before passing it on to the next.

At most stages some HYDROGEN or CARBON DIOXIDE is liberated from the molecule.

GLUCOSE

PYRUVATE

when Pyruvate is oxidized the released energy is not lost as heat – but is used to form a high energy compound—Acetylcoenzyme A.

Acetylcoenzyme A

Oxaloacetate — 2H

H_2O Malate

Citrate H_2O

H_2O Fumarate

cis-aconitate   — 2H

H_2O Succinate

Isocitrate

2H H_3PO_4

CO_2 α-oxoglutarate   Succinylcoenzyme A

2H

CO_2

The liberated HYDROGEN

is transported by CARRIERS

Molecular OXYGEN is supplied by blood

| CYTOCHROME SYSTEM | 2H | FLAVOPROTEINS | 2H | NICOTINAMIDE NUCLEOTIDES |

H_2O with liberation of **ENERGY**

In the body this ENERGY, instead of being lost as heat, is used to link an additional phosphate group to an existing molecule of ADP (adenosine diphosphate) to form the energy-rich compound ATP These compounds are stored and the energy 'trapped' in them is used as required ——

e.g.

to _build_ more complex STORAGE, STRUCTURAL or FUNCTIONAL materials such as Protein of new tissue in growth and repair, Secretions, Enzymes, etc.

for _work of the cell_ — in ways not fully understood — e.g. for muscle contraction and transmission of nerve impulse.

to _liberate heat_ to keep body warm.

This KREBS CITRIC-ACID CYCLE is the chief pathway by which potential energy stored in food is released by cells to make their own energy-rich Phosphorus compound – ADENOSINE TRIPHOSPHATE (ATP).

HEAT BALANCE

ENERGY is released in cells by OXIDATION. It is used for work and to keep body warm. The Body Temperature is kept relatively constant (*with a slight fluctuation throughout the 24 hours*) in spite of wide variations in Environmental Temperature and Heat Production.

HEAT PRODUCTION ──────── *must balance* ──────── **HEAT LOSS**

depends on kind and amount of Food eaten. About 95% of the ENERGY released by OXIDATION of FOODSTUFFS appears in the body as HEAT.

e.g.   **ENERGY INTAKE**
in
FOOD
of 3300 kilocalories

released by OXIDATION in cells

| 300 kcal UTILIZED as WORK by | 3000 kcal APPEAR as a by-product of metabolism – HEAT from |

ALL METABOLIZING CELLS but especially **ACTIVE TISSUES**

such as

LIVER

SECRETING GLANDS and

MUSCLES

This HEAT is rapidly dispersed and transported throughout the Body by BLOOD to TISSUE FLUIDS

from GASTRO-INTESTINAL TRACT

Warming ingested COLD FLUID and FOOD

from LUNGS

Warming INSPIRED COLD AIR

Loss in VAPORIZATION of H_2O in EXPIRED AIR

110 kcal   240 kcal

from TISSUE FLUIDS

BLOOD

from URINARY TRACT

Loss of Heat in FAECES   URINE 50 kcal

from SKIN SURFACE
by RADIATION CONVECTION CONDUCTION
2100 kcal
Amount lost depends on calibre of skin blood v's and amount of blood flow.

EVAPORATION of WATER

1. From lungs
2. Insensible Perspiration (Osmosis: Diffusion) 500 kcal
3. Secretion of Sweat

SKIN

[*Only effective for Heat loss if allowed to evaporate on Body Surface*] (Every Gram takes 0.58 kcal to EVAPORATE and AMOUNT VARIES)

43

MAINTENANCE of BODY TEMPERATURE

Any tendency for the BODY TEMPERATURE to *RISE*
as by

INCREASED HEAT PRODUCTION is balanced by **INCREASED HEAT LOSS**

1. by increased cellular
   <u>OXIDATION of FOODSTUFFS</u>
   as occurs e.g. with
   <u>MUSCULAR ACTIVITY</u>

   EXTRA HEAT
   is dispersed
   quickly by Blood Stream

   Rise in Blood
   Temperature
   affects
   <u>HYPOTHALAMUS</u>

2. <u>HOT ENVIRONMENTAL
   TEMPERATURE</u>
   i.e. above BODY
   TEMPERATURE (37°C)
   *[Body would tend
   to gain heat by
   RADIATION,
   CONVECTION and
   CONDUCTION.]*

   stimulates

   <u>HEAT SENSITIVE</u>
   nerve endings
   in SKIN

   INGOING
   NERVE
   IMPULSES

*Reduced
SYMPATHETIC
VASOCONSTRICTOR
tone to*
SECRETOMOTOR

BLOOD
VESSELS
DILATED

FAT

OUTGOING
NERVE IMPULSES

1. SKIN BLOOD VESSELS DILATE

   *more blood to skin
   surface → increased heat
   loss from skin by —*
   RADIATION
   CONVECTION   *(Cannot
   CONDUCTION   occur if air
                 temperature
                 is above
                 body's)*
   *Increased by
   voluntary ingestion of
   cold foods and fluids
   and by use of fans.*

2. SWEAT GLANDS SECRETE
   *Increased heat loss by
   evaporation from skin
   surface (unless atmosphere
   is already water saturated
   as e.g. in Tropics)*

3. <u>DIMINISHED HEAT
   INSULATION</u>
   *by voluntary reduction
   of CLOTHING worn*

4. <u>DIMINISHED HEAT
   PRODUCTION</u>
   SKELETAL MUSCLE *'tone'
   reduced – and often
   voluntary relaxation
   → less work done →
   less heat produced*

5. <u>REDUCTION of 'ENERGY INTAKE'</u>
   *by voluntary restriction of
   <u>PROTEIN in DIET</u>*

*[The increase in activity during the day
probably accounts for the gradual Physiological
rise in body temperature from about 96·5°F
(35·8°C) in the early morning to about 99·2°F
(37·3°C) in the late afternoon.*

*Unless exercise is very strenuous or environ-
ment is very hot and humid these measures →* RESTORE BODY TEMPERATURE
to NORMAL

MAINTENANCE of BODY TEMPERATURE

Any tendency for the BODY TEMPERATURE to *FALL*
as by

DIMINISHED HEAT PRODUCTION is balanced by **DIMINISHED HEAT LOSS**

1. by decreased cellular
   OXIDATIONS as during
   MUSCULAR INACTIVITY
   (e.g. during sleep)

   LESS HEAT
   transported by Blood Stream

2. or
   COLD ENVIRONMENTAL
   TEMPERATURE
   i.e. below CRITICAL
   TEMPERATURE
   *(about 30°C. for
   naked body)*

   stimulates

   COLD SENSITIVE
   nerve endings
   in SKIN

   INGOING
   NERVE
   IMPULSES

*Drop in Blood
Temperature
affects*
HYPOTHALAMUS

Increased
SYMPATHETIC
VASOCONSTRICTOR
tone to →

SUPPRESSION

OUTGOING
NERVE IMPULSES

*Suprarenal
Medulla*

*Smooth
Muscle
and
Blood
Vessels*

Adrenaline

BLOOD
VESSELS
CONSTRICTED

FAT

1. SKIN BLOOD VESSELS CONSTRICT

   *less blood flowing to skin
   surface → less heat loss
   from skin by –*
   RADIATION
   CONVECTION }
   CONDUCTION

   *Man reduces these losses
   still further by intake of
   warm food, fluids and by
   heating environment.*

2. SWEAT GLANDS SUPRESSED

   *reduced heat loss from
   evaporation of water.*

3. SUBCUTANEOUS FAT –
   *an insulating layer
   through which heat
   passes with difficulty.*
   INCREASED HEAT
   INSULATION
   *by voluntary use of
   CLOTHING*

   SMOOTH MUSCLE *of skin
   contracts (Gooseflesh")*

4. ↑INCREASED HEAT
   PRODUCTION
   SKELETAL MUSCLE *shows
   involuntary increased 'tone'-
   "shivering". Voluntary activity
   → more work done →
   more heat produced.*

   *In very cold climates
   there is a tendency to*

5. INCREASE 'ENERGY INTAKE'
   *by voluntary increase in
   DIETARY PROTEIN which has
   stimulating effect on metabolism.*

[*NOTE: In temperate climates
environmental temperature is
usually lower than body temperature
so that there is a continuous loss of
heat from body surface.]*

*Unless environmental temperature
is very low these measures tend to* — RESTORE BODY TEMPERATURE
to NORMAL

GROWTH

Each individual grows, by repeated cell divisions, from a single cell to a total of 30 trillion or more cells. Growth is most rapid before birth and during 1ST year of life.

The proportion of ENERGY INTAKE in food used to build and maintain tissue —	INFANCY		CHILDHOOD		
	At Birth	3 months	1 year	2 years	9-11 years
	40%	40%	20%	20%	4-10%
Average WEIGHT —	3 kg	5 kg	10·4 kg	12·4 kg	27·1 kg
Average HEIGHT —	50 cm	58 cm	73 cm	84 cm	129 cm

A baby is born with epithelia, connective tissues, muscles, nerves and organs all present and formed — but all tissues do not grow at the same rate.

DIFFERENTIAL GROWTH and FUNCTIONAL DEVELOPMENT of TISSUES

e.g. Rapid growth of skeletal tissue during childhood.
Nervous tissue develops rapidly in first 2 years.
Most rapid growth is first at the head then legs begin to lengthen.

↓ *lead to*

CHANGE in BODY PROPORTIONS

Chiefly PROTEIN being laid down →
or retained

Lymphoid Tissue Growth Spurt
10 YEARS

2 YEARS

1 YEAR

3 MONTHS

At BIRTH

FIRST *(Neutral)* **GROWTH PHASE** *(Infancy to Puberty)*. No marked difference between sexes – Regulated by Growth-Stimulating Hormone of Anterior Pituitary – stimulates Growth of all Tissues and Organs.

FACTORS INFLUENCING NORMAL GROWTH :
HEREDITY: *To large extent rate of growth/sequence of events is predetermined. Inherited factors control pattern and limitations of growth.*
ENVIRONMENT: *Nutrition: For optimal growth the body requires an adequate and balanced diet.*
ENDOCRINE GLANDS: *All play part: mental as well as physical development is influenced.*

GROWTH

GROWTH SPURT 11-12 years	ADOLESCENCE →		ADULTHOOD →	
	GIRL	**BOY**	**WOMAN**	**MAN**
	13 years	*15 years*	*17 years*	*19 years*
	10-15%	10-15%	4%	4%
	42·8 kg	51·1 kg	54·8 kg	65 kg
	152 cm	163 cm	162 cm	173 cm

Growth and development of reproductive organs

More FAT being deposited now

A part from PROTEIN replacement in REPAIR, weight increase is now by deposition of FAT.

SEXUAL GROWTH PHASE *(Puberty to Maturity)*

Marked difference between sexes is initiated by Gonad-Stimulating Hormones of Anterior Pituitary acting on Sex Glands and stimulating their production of Sex Hormones. These hormones are largely responsible for development of secondary sex characteristics and development of reproductive organs.

REPRODUCTIVE PHASE

↓

These hormones maintain secondary sex characteristics and reproductive ability during reproductive phase of adult life.

47

ENERGY REQUIREMENTS *MALE*

DAILY FOOD INTAKE must supply <u>TOTAL ENERGY REQUIREMENTS</u> for <u>1 ACTIVITY</u>

SPECIAL – Individual requirements vary with TYPE of WORK or PLAY and, from minute to minute, on the INTENSITY of WORK and frequency and length of REST PAUSES.

EVERYDAY ACTIVITIES– such as SITTING, STANDING, WALKING, ETC.

2. SPECIFIC DYNAMIC ACTION of FOOD (S.D.A.)

The mere taking of food stimulates metabolism of cells so that heat production increases (30% by protein, 4-5% by carbohydrate and fat). Must allow 10% above Basal Requirements on average mixed diet.

3. BASAL METABOLISM

(measured when body is at rest) Energy Expenditure of cells doing 'vegetative' processes of living – e.g. tasks involved in RESPIRATION, CIRCULATION, DIGESTION, EXCRETION, SECRETION, SYNTHESIS of special substances, KEEPING BODY TEMPERATURE at 37°C., GROWTH and REPAIR.

Average 70 kg 25 year old ADULT (surface area 1·8 square metres)

ADOLESCENT 16 years

CHILD 8 years

HEAVY WORK   MODERATE   LIGHT (Sedentary)

Very active in play and everyday activities

HEAVY WORK
4500 kcal

Up to 380 kcal per hour at peak performance. 2240 kcal used in average 8 hour day.

3380 kcal

Up to 140 kcal per hour. 1120 kcal/day.

2420 kcal

e.g. 160 kcal/day.

3500 kcal

e.g. 1560 kcal/day.

2000 kcal

e.g. 743 kcal/day.

EVERYDAY ACTIVITY 360 kcal/day.

S.D.A. 173 kcal/day.

176

114

BASAL METABOLISM 1728 kcal/day

1728

1728

1764

1143

40 kcal/square metre/hour

45·7 kcal/m²/hour

51·8 kcal/m²/hour

Basal metabolism is proportional to surface area of body and varies with sex and age.

[All figures are approximate and intended only as a general guide.]

ENERGY REQUIREMENTS *FEMALE*

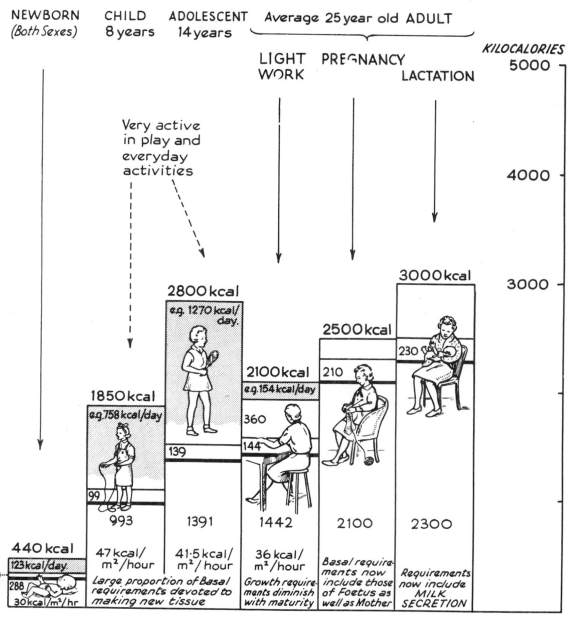

NEWBORN (Both Sexes)   **CHILD** 8 years   **ADOLESCENT** 14 years   Average 25 year old **ADULT**

LIGHT WORK   **PREGNANCY**   **LACTATION**

KILOCALORIES
5000

Very active in play and everyday activities

4000

3000kcal
3000

2800kcal
e.g. 1270 kcal/day.

2500kcal

2100kcal
e.g. 154 kcal/day
230

1850kcal
e.g.758 kcal/day
210

360

139
144
99

993   1391   1442   2100   2300

440 kcal
123 kcal/day
29
288
30 kcal/m²/hr

47 kcal/ m²/hour
Large proportion of Basal requirements devoted to making new tissue

41.5 kcal/ m²/hour

36 kcal/ m²/hour
Growth requirements diminish with maturity

Basal requirements now include those of Foetus as well as Mother

Requirements now include MILK SECRETION

[Proportion needed for growth diminishes in both sexes with age.]

49

BALANCED DIET

The individual's daily energy requirements are best obtained by eating well-balanced meals which contain the essential *body-building* and *protective* foods as well as *energy-giving* foods.

1st. Class Proteins

Milk & Milk Foods

TINNED MILK

Vit.B₂ Ca P

Vit.D Nacl

FROM EACH GROUP:-

Children:
3-4 servings per day.
Adults:
2 or more servings per day.

1st. Class Proteins

Ca
Vit. B₂
Vit.D Iodine
Fe NaCl S K

1-2 servings per day.

Adult total protein requirement averages 70 to 100 grams/day.

2nd. Class Proteins
(with some carbohydrates)

Vit.C FROZEN CORN FOOD

Fe Vit.C Vit.B I₂ K S Vit.A

1 serving per day

1 serving per day

NOTE:- The normal daily requirement of vitamins and essential minerals is met by such a balanced diet. In addition to fluid in these foods, daily <u>Water</u> *requirement is approximately 1 litre. This varies with sweat loss, etc. (see Index under "Water Balance")*

To Provide chiefly Vitamin

ROSE HIP SYRUP

C

Chiefly, for their carbohydrates

CORN FLAKES Vit.B Fe

JAM SUGAR

(Vit.C)

4 or more servings per day

For Fats and Fat Soluble Vitamins

PEANUT BUTTER COOKING FAT MARG

(NaCl)

Some per day

50

CHAPTER 3

DIGESTIVE SYSTEM

DIGESTIVE SYSTEM

The ALIMENTARY CANAL and
ASSOCIATED GLANDS } Special System for dealing
with FOOD and FLUIDS

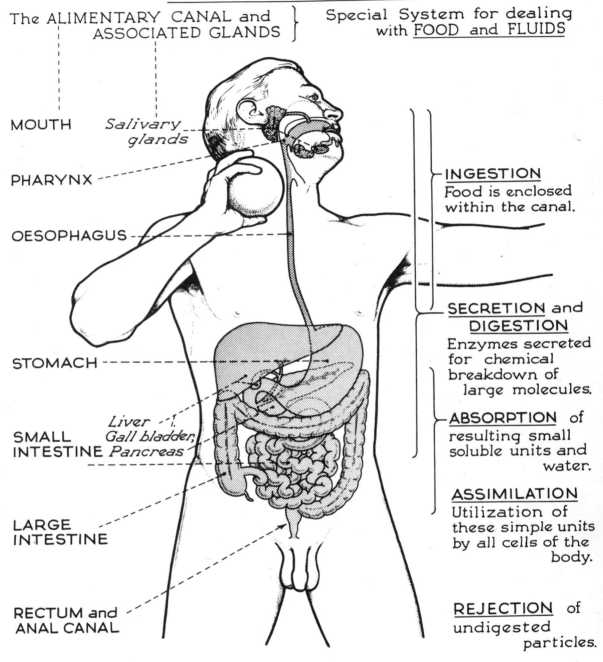

MOUTH   *Salivary glands*

PHARYNX

OESOPHAGUS

STOMACH

SMALL INTESTINE   *Liver* *Gall bladder,* *Pancreas*

LARGE INTESTINE

RECTUM and ANAL CANAL

INGESTION
Food is enclosed
within the canal.

SECRETION and DIGESTION
Enzymes secreted
for chemical
breakdown of
large molecules.

ABSORPTION of
resulting small
soluble units and
water.

ASSIMILATION
Utilization of
these simple units
by all cells of the
body.

REJECTION of
undigested
particles.

PROGRESS of FOOD along ALIMENTARY CANAL

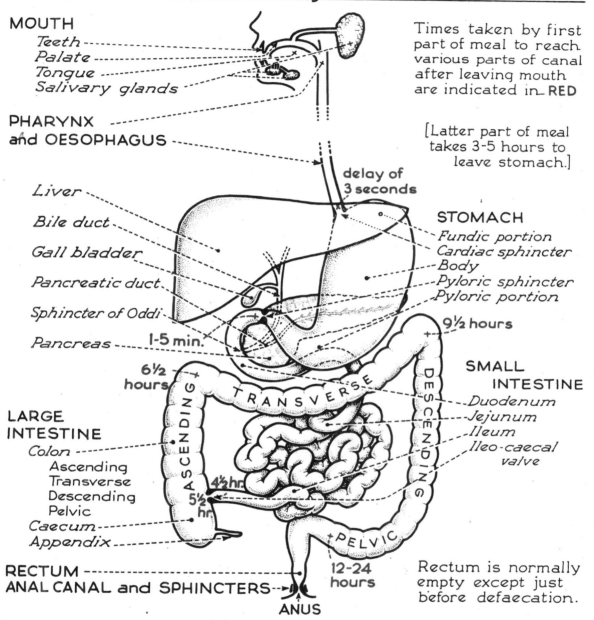

MOUTH
  Teeth
  Palate
  Tongue
  Salivary glands

PHARYNX
and OESOPHAGUS

Times taken by first part of meal to reach various parts of canal after leaving mouth are indicated in **RED**

[Latter part of meal takes 3-5 hours to leave stomach.]

delay of 3 seconds

Liver

Bile duct

Gall bladder

Pancreatic duct

Sphincter of Oddi

Pancreas — 1-5 min.

STOMACH
  Fundic portion
  Cardiac sphincter
  Body
  Pyloric sphincter
  Pyloric portion

9½ hours

6½ hours

SMALL INTESTINE
  Duodenum
  Jejunum
  Ileum
  Ileo-caecal valve

TRANSVERSE

ASCENDING

DESCENDING

LARGE INTESTINE
  Colon
    Ascending
    Transverse
    Descending
    Pelvic
  Caecum
  Appendix

4½ hr.

5½ hr.

PELVIC

RECTUM
ANAL CANAL and SPHINCTERS

12-24 hours

ANUS

Rectum is normally empty except just before defaecation.

During its progress along the canal FOOD is subjected to MECHANICAL as well as CHEMICAL changes to render it suitable for absorption and assimilation.

DIGESTION in the MOUTH

MECHANICAL PROCESSES

MASTICATION

Chewing movements of TEETH, TONGUE, CHEEKS, LIPS, LOWER JAW, break down food, mix it with SALIVA and roll it into a moist soft mass (BOLUS) suitable for SWALLOWING

[Mastication is under VOLUNTARY control]

PAROTID SALIVARY GLAND *(serous)*

SUBLINGUAL and SUBMANDIBULAR SALIVARY GLANDS *(mixed mucous and serous glands)*

CHEMICAL PROCESSES

SALIVA *(1-1½ litres per day)* is a slightly acid solution of salts and organic substances secreted mainly by 3 pairs of SALIVARY GLANDS.

SEROUS ACINI → Clear watery fluid containing salts and PTYALIN — an enzyme which starts to split cooked STARCH → DEXTRINS → MALTOSE

WATER
Solid substances must be dissolved in saliva to stimulate TASTE buds.

MUCOUS ACINI → Thick slimy secretion of MUCIN → lubricant coating to food to assist SWALLOWING

Crescent of Serous cells

[Salivary secretion is REFLEX and INVOLUNTARY]

Other important *(non-digestive)* functions of SALIVA :—

CLEANSING — Mouth and teeth kept free of debris, etc., to inhibit bacteria.

MOISTENING and LUBRICATING — Soft parts of mouth kept pliable for SPEECH.

EXCRETORY — Many organic substances *(e.g. urea, sugar)* and inorganic substances *(e.g. mercury, lead)* can be excreted in saliva.

CONTROL of SALIVARY SECRETION

Increased secretion at mealtimes is REFLEX *(involuntary)*.

Salivary Reflexes are of two types:–

(a) <u>UNCONDITIONED</u> *(inborn)*

INGOING messages along TASTE nerves to Controlling SALIVARY Centre in MEDULLA OBLONGATA

OUTGOING messages in SECRETOMOTOR nerves from the AUTONOMIC NERVOUS SYSTEM Centre to glands evoke secretion of SALIVA.

FOOD stimulates nerve endings in mouth

(b) <u>CONDITIONED</u> *(depend on experience)*

SIGHT of food becomes associated with TASTE of food

INGOING messages to VISUAL Centre

INGOING messages to SALIVARY Centre

ASSOCIATION pathways become established between centres in the Brain

Eventually SIGHT of food alone.....

..... evokes SALIVARY secretion

Similar conditioned reflexes are established by smell, by thought of food, and even by the sounds of its preparation.

OESOPHAGUS

The Oesophagus is a muscular tube about 25 cm. long which conveys ingested food and fluid from the Mouth to the Stomach.

PHARYNX

The Muscle is arranged in 2 layers:-
Outer LONGITUDINAL
Inner CIRCULAR

In the UPPER THIRD of the tube the muscle is SKELETAL

In the MIDDLE THIRD the muscle is mixed SKELETAL and VISCERAL

In the LOWER THIRD the muscle is SMOOTH, VISCERAL or INVOLUNTARY

×15

×15

×15

Relaxation of Cardiac Sphincter permits food to enter
DIAPHRAGM
STOMACH

The tube has 4 Coats:-

1. MUCOSA (lining)
Stratified Squamous Epithelium (non-keratinized) -a thick protective surface layer.

Muscularis Mucosae — Smooth muscle of the Mucosa.

Mucous Glands in the
2. SUBMUCOSA secrete a viscid lubricant mucus which passes along ducts to the inner surface of the tube.

3. MUSCLE COATS
Contraction of these occurs in swallowing.

Between the two Muscle Coats lie the nerves and nerve cells of an Intrinsic Nerve Plexus (Auerbach). Through this network PERISTALTIC MOVEMENTS of the Muscle Coats are controlled.

4. An outer coat of CONNECTIVE TISSUE blends with that of trachea, etc.

Except during passage of food the oesophagus is flattened and closed; its mucosa thrown into several longitudinal folds.

SWALLOWING

Swallowing is a complex act initiated *voluntarily* and completed
involuntarily (or reflexly)

STIMULUS	INGOING PATHWAY	CENTRE IN MEDULLA OBLONGATA	OUTGOING PATHWAY	EFFECT

BUCCAL STAGE
(Voluntary)

FOOD BOLUS stimulates nerve endings at back of TONGUE and THROAT →

in Cranial nerves IX, X

NASO-PHARYNX
HARD PALATE
BOLUS
TONGUE
PHARYNX
LARYNX (AIR PASSAGE TO LUNGS)

Cranial nerves *IX, X, XII* to excite muscles at base of tongue and floor of mouth and in Phrenic and Intercostal nerves to inhibit diaphragm and intercostal muscles.

Tongue is drawn upwards and backwards against hard palate.

Food bolus is projected into pharynx.

Respiration is inhibited.

PHARYNGEAL STAGE
(Involuntary)

FOOD BOLUS stimulates nerve endings in mucosal lining of PHARYNX →

in Cranial nerve X

SOFT PALATE

Cranial nerves *IX, XI* to excite muscles of pharyngeal wall and muscles of soft palate, laryngeal wall and base of tongue.

Pharyngeal wall contracts, grips and squeezes bolus into oesophagus.

Soft palate is raised to prevent food entering nose.

Laryngeal wall is raised to safeguard bronchial passages.

Tongue is drawn still further back.

OESOPHAGEAL STAGE
(Involuntary)

FOOD BOLUS stimulates nerve endings in wall of OESOPHAGUS →

Intrinsic Nerve Plexus

OESOPHAGUS
CARDIAC SPHINCTER
STOMACH

AUERBACH'S (intrinsic) *Plexus* to excite muscle coats of oesophagus and in *Vagus nerve* to inhibit muscles of Cardiac sphincter

Peristaltic waves of contraction squeeze bolus towards cardiac sphincter.

Cardiac sphincter relaxes and allows bolus to pass into stomach.

STOMACH

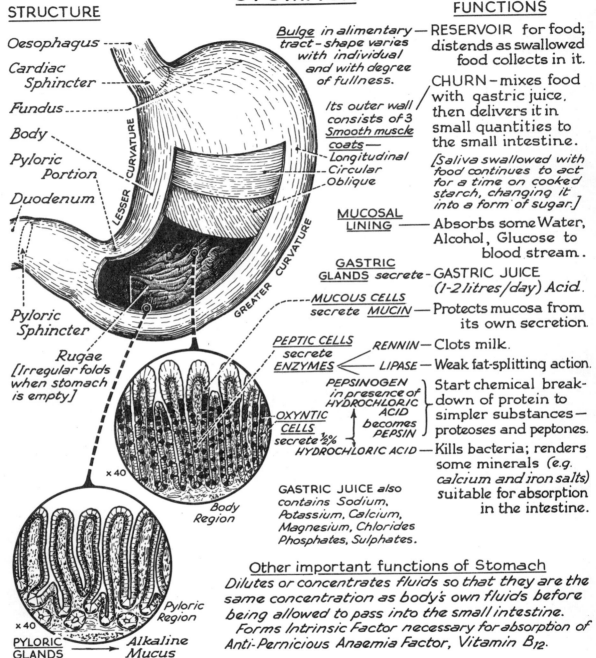

STRUCTURE

Oesophagus

Cardiac Sphincter

Fundus

Body

Pyloric Portion

Duodenum

LESSER CURVATURE

GREATER CURVATURE

Pyloric Sphincter

Rugae
[Irregular folds when stomach is empty]

× 40

× 40

PYLORIC GLANDS

Body Region

Pyloric Region

Alkaline Mucus

Bulge in alimentary tract – shape varies with individual and with degree of fullness.

Its outer wall consists of 3 Smooth muscle coats —
- Longitudinal
- Circular
- Oblique

MUCOSAL LINING

GASTRIC GLANDS secrete

MUCOUS CELLS secrete MUCIN

PEPTIC CELLS secrete ENZYMES — RENNIN
— LIPASE

PEPSINOGEN in presence of HYDROCHLORIC ACID becomes PEPSIN

OXYNTIC CELLS secrete ½% HYDROCHLORIC ACID

GASTRIC JUICE also contains Sodium, Potassium, Calcium, Magnesium, Chlorides Phosphates, Sulphates.

FUNCTIONS

RESERVOIR for food; distends as swallowed food collects in it.

CHURN — mixes food with gastric juice, then delivers it in small quantities to the small intestine.

[Saliva swallowed with food continues to act for a time on cooked starch, changing it into a form of sugar.]

Absorbs some Water, Alcohol, Glucose to blood stream.

GASTRIC JUICE (1-2 litres/day) Acid.

Protects mucosa from its own secretion.

Clots milk.

Weak fat-splitting action.

Start chemical break-down of protein to simpler substances — proteoses and peptones.

Kills bacteria; renders some minerals (e.g. calcium and iron salts) suitable for absorption in the intestine.

Other important functions of Stomach

Dilutes or concentrates fluids so that they are the same concentration as body's own fluids before being allowed to pass into the small intestine.

Forms Intrinsic Factor necessary for absorption of Anti-Pernicious Anaemia Factor, Vitamin B_{12}.

GASTRIC JUICE

SECRETION of Gastric Juice is under 2 types of CONTROL:-

(a) NERVOUS - Messages are conveyed rapidly from Brain Centre by Nerve Fibres of the AUTONOMIC NERVOUS SYSTEM for *immediate* effect. e.g. stimulation of Parasympathetic (Vagus) nerves to GASTRIC GLANDS → secretion of HIGHLY ACID JUICE containing ENZYMES.

(b) HUMORAL - Message is CHEMICAL and carried in BLOOD STREAM for *slower* and *longer-lasting* control. [Note:- *Chemical messengers (Hormones) travel all over the body even though the message stimulates only one part — e.g. in this case the Gastric Glands.*]

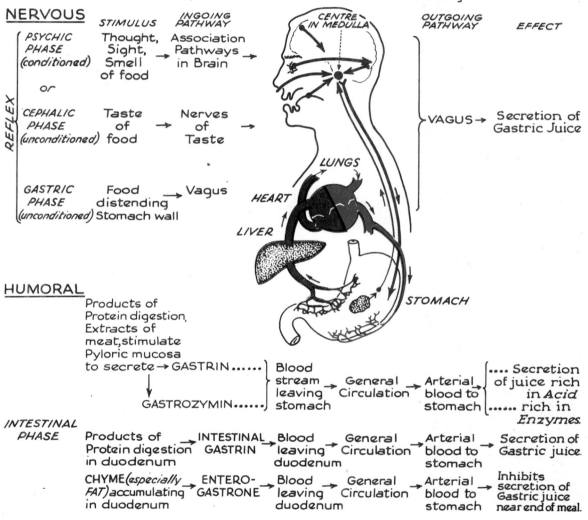

NERVOUS

	STIMULUS	INGOING PATHWAY			OUTGOING PATHWAY	EFFECT
PSYCHIC PHASE (conditioned)	Thought, Sight, Smell of food	Association Pathways in Brain →				
or						
CEPHALIC PHASE (unconditioned)	Taste of food	Nerves of Taste →	→		VAGUS →	Secretion of Gastric Juice
GASTRIC PHASE (unconditioned)	Food distending Stomach wall	→ Vagus				

REFLEX

CENTRE IN MEDULLA

LUNGS

HEART

LIVER

STOMACH

HUMORAL

Products of Protein digestion, Extracts of meat, stimulate Pyloric mucosa to secrete → GASTRIN ...... ⎱ Blood stream leaving stomach → General Circulation → Arterial blood to stomach ⎱ .... Secretion of juice rich in *Acid*

                            ↓
            GASTROZYMIN ...... ⎰   ...... rich in *Enzymes.*

INTESTINAL PHASE

Products of Protein digestion in duodenum → INTESTINAL GASTRIN → Blood leaving duodenum → General Circulation → Arterial blood to stomach → Secretion of Gastric juice.

CHYME *(especially FAT)* accumulating in duodenum → ENTERO-GASTRONE → Blood leaving duodenum → General Circulation → Arterial blood to stomach → Inhibits secretion of Gastric juice near end of meal.

MOVEMENTS of the STOMACH

Very little movement is seen in empty stomach until onset of HUNGER.

FILLING

FUNDUS and
GREATER CURVATURE bulge and lengthen as stomach fills with food.   This is "Receptive Relaxation" – the smooth muscle cells relax so that the pressure in the stomach does not rise until the organ is fairly large.

TWO TYPES of MOVEMENT occur while food is in stomach.

TONUS

HOPPER

BODY – Food lies in layers and the walls exert a slight but steady pressure on it. This squeezes food steadily towards the PYLORUS even when stomach is becoming relatively empty near the end of a meal.

PERISTALSIS

MILL

PYLORIC
SPHINCTER

From INCISURA ANGULARIS vigorous waves of contraction mix food with digestive juices and carry chyme through the normally relaxed PYLORUS into the first part of the duodenum.

EMPTYING
CONTROL of MOTILITY and EMPTYING

1. Enterogastric
   NERVOUS Reflex   *inhibition of VAGUS →*   Inhibits peristalsis temporarily.
     As Chyme enters and   ↓
     distends duodenum   Gastric motility depressed.
   HUMORAL   *release of ENTEROGASTRONE →*   Temporarily slows emptying
                   to Blood Stream   of stomach.

2.
                      stimulation of VAGUS →   Waves of peristalsis become
     As duodenum   stronger.
         empties   Gastric motility increased.
                 withdrawal of ENTEROGASTRONE → Emptying speeded up again.

The hormone is particularly important in regulating the rate of emptying of the stomach from moment to moment during the digestion of a meal.

[STARCHY FOODS leave the stomach quickly: MEAT leaves relatively slowly: FATTY FOODS pass through most slowly of all.]

60

VOMITING

Vomiting is a <u>REFLEX</u> act.

Nerve pathways involved:-

STIMULUS → *INGOING NERVE PATHWAY* → *VOMITING CENTRE IN MEDULLA OBLONGATA* → *OUTGOING NERVE PATHWAY* → *EFFECT*

<u>PSYCHIC INFLUENCES</u>
e.g. fear, anxiety, disgust.

<u>DISCORDANT INGOING IMPULSES</u>
    <u>EYES</u>
    <u>LABYRINTH of EARS</u>
    e.g. stimulation of nerve
    endings of utricle in
    Motion Sickness.

<u>LOCAL IRRITATION</u> of
    nerve endings
    (parasympathetic and sympathetic)
    by drugs, poisons, or in
    disease in PHARYNX,
    OESOPHAGUS, STOMACH,
    INTESTINE *(especially*
    APPENDIX), GALL
    BLADDER, UTERUS, in
    FAILING HEART or from
    pain endings in any
    organ of the body.

<u>BLOOD BORNE EFFECTS</u>
    Certain EMETIC DRUGS,
    e.g. *Apomorphine,* lower
    threshold of Centre
    to stimulation.
    <u>METABOLIC</u>
        <u>DISTURBANCES</u>
    e.g. in Pregnancy and
        Fatigue

VAGUS — *Oesophagus*
      — *Cardiac*
        sphincter ⎫→ Relax.
      — *Body of*
        stomach ⎭
      — *Pyloric* → Contracts
        part   strongly.

SPINAL — *Abdominal*
NERVES   *muscles* → Contract.

PHRENIC — *Diaphragm* →
NERVE   Descends to
         compress
         abominal cavity.

*[Larynx is raised to
protect air passages.]*

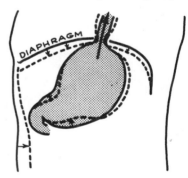

DIAPHRAGM

As a result of these changes
the <u>Stomach is compressed</u>
with <u>PASSIVE emptying of
contents.</u>

PANCREAS

The Pancreas is a large gland lying across the Posterior Abdominal Wall. It has 2 secretions – a *digestive* secretion poured into the duodenum, and a *hormonal* secretion passed into the blood stream.

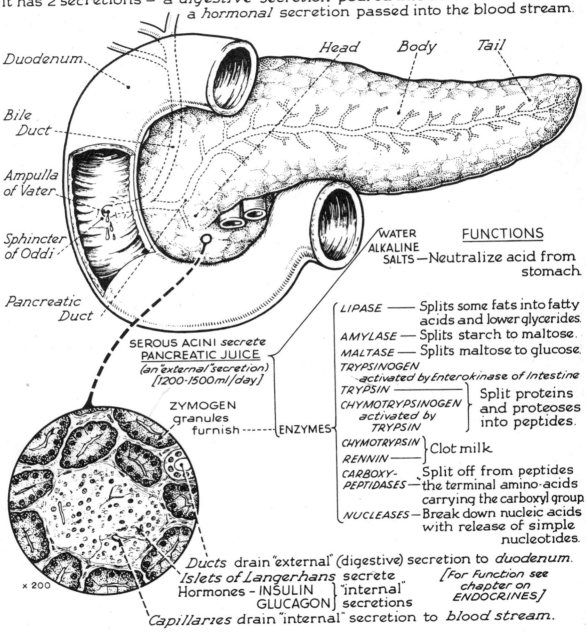

Duodenum

Bile Duct

Ampulla of Vater

Sphincter of Oddi

Pancreatic Duct

Head   Body   Tail

WATER
ALKALINE
SALTS

×200

SEROUS ACINI *secrete*
PANCREATIC JUICE
(an "external" secretion)
[1200-1500ml/day]

ZYMOGEN
granules
furnish ------ ENZYMES

FUNCTIONS

WATER ALKALINE SALTS — Neutralize acid from stomach.

LIPASE ——— Splits some fats into fatty acids and lower glycerides.

AMYLASE — Splits starch to maltose.

MALTASE — Splits maltose to glucose.

TRYPSINOGEN activated by Enterokinase of Intestine

TRYPSIN ———
CHYMOTRYPSINOGEN activated by *TRYPSIN*
} Split proteins and proteoses into peptides.

CHYMOTRYPSIN
RENNIN ———
} Clot milk

CARBOXY-PEPTIDASES — Split off from peptides the terminal amino-acids carrying the carboxyl group.

NUCLEASES — Break down nucleic acids with release of simple nucleotides.

Ducts drain "external" (digestive) secretion to *duodenum*.

Islets of Langerhans *secrete*
Hormones – INSULIN GLUCAGON } "internal" secretions

[For Function see chapter on ENDOCRINES]

Capillaries drain "internal" secretion to *blood stream*.

62

PANCREATIC JUICE

SECRETION of Pancreatic Juice is under 2 types of CONTROL:-
NERVOUS and HUMORAL.

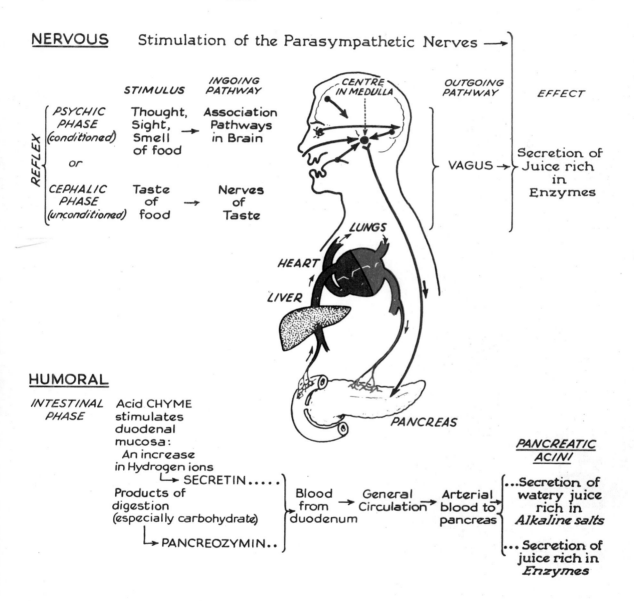

NERVOUS   Stimulation of the Parasympathetic Nerves ⟶

	STIMULUS	INGOING PATHWAY		OUTGOING PATHWAY	EFFECT
PSYCHIC PHASE (conditioned)	Thought, Sight, Smell of food	Association Pathways in Brain		VAGUS ⟶	Secretion of Juice rich in Enzymes
or					
CEPHALIC PHASE (unconditioned)	Taste of food	Nerves of Taste			

CENTRE IN MEDULLA

LUNGS

HEART

LIVER

PANCREAS

HUMORAL

INTESTINAL PHASE   Acid CHYME stimulates duodenal mucosa:
   An increase in Hydrogen ions
      ⟶ SECRETIN.....
   Products of digestion (especially carbohydrate)
      ⟶ PANCREOZYMIN..

Blood from duodenum ⟶ General Circulation ⟶ Arterial blood to pancreas

PANCREATIC ACINI

...Secretion of watery juice rich in *Alkaline salts*

...Secretion of juice rich in *Enzymes*

LIVER and GALL BLADDER

The Liver is a large highly complex organ with many functions. One of these is the production of BILE (500-1000 ml/day)

Fibrous tissue Capsule

Hepatic Ducts

Liver Lobule

Cystic Duct

Central Vein

Common Bile Duct

Portal Canal with branch of Hepatic Artery, Portal Vein, Bile Duct

Gall Bladder

DUODENUM

Plates of Liver Cells Bile Canaliculi

Blood Sinusoids

Kupffer Cell

LIVER CELLS – secrete- BILE (slightly alkaline) contains Lecithin, Cholesterol, Inorganic Salts

–unite CHOLIC ACID with
GLYCINE → SODIUM GLYCOCHOLATE
TAURINE → SODIUM TAUROCHOLATE } → BILE SALTS

– convert breakdown products of HAEMOGLOBIN → BILE PIGMENTS (Biliverdin → Bilirubin)

– store IRON

– discharge BILE into BILE CANALICULI

System of Channels

HEPATIC DUCT ⟶ STORAGE in Gall Bladder (up to 50cc.)

GALL BLADDER

Columnar epithelium → Mucus added.

H_2O and Salts reabsorbed → CONCENTRATION (x 10).

Contraction of Smooth Muscle → EXPULSION to duodenum.

In INTESTINE ——— FUNCTIONS of BILE

BILE SALTS (with fatty acids and glycerides) → stable emulsion of fats → Facilitates action of LIPASE

Activate Pancreatic LIPASE

Form water-soluble complexes with some water-insoluble substances (e.g. Cholesterol; fat-soluble Vitamins)

PROMOTE their ABSORPTION

small amount ——— most reabsorbed ——— LIVER stimulate further excretion of Bile (CHOLERETIC ACTION)

BILE PIGMENTS ——— some reabsorbed ——— LIVER
Bacterial action
STERCOBILINOGEN

Re-excretion in Bile
Systemic Circulation
KIDNEY
UROBILINOGEN

STERCOBILIN (Brown pigment of Faeces)

UROBILIN of URINE

Other substances, e.g. certain drugs, can be taken out of the blood stream by the LIVER CELLS and EXCRETED in the BILE.

EXPULSION of BILE *(from the Gall Bladder)*

BILE is SECRETED continuously by the LIVER.
It is STORED and CONCENTRATED in the GALL BLADDER.
Periodically *(e.g. during a meal)* it is DISCHARGED into the DUODENUM.
NERVOUS and HUMORAL factors influence this expulsion:-

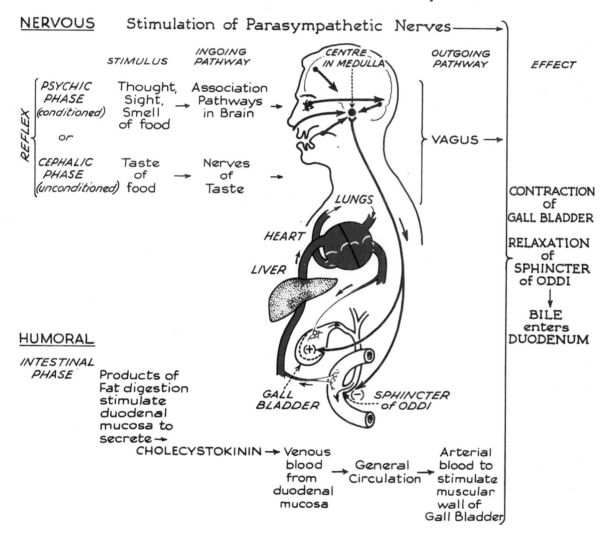

NERVOUS   Stimulation of Parasympathetic Nerves ──────→

	STIMULUS	INGOING PATHWAY	CENTRE IN MEDULLA	OUTGOING PATHWAY	EFFECT

REFLEX —
PSYCHIC PHASE (conditioned) — Thought, Sight, Smell of food → Association Pathways in Brain →

or

CEPHALIC PHASE (unconditioned) — Taste of food → Nerves of Taste →

VAGUS →

LUNGS

HEART

LIVER

CONTRACTION
of
GALL BLADDER

RELAXATION
of
SPHINCTER
of ODDI
↓
BILE
enters
DUODENUM

HUMORAL

INTESTINAL PHASE — Products of Fat digestion stimulate duodenal mucosa to secrete →

GALL BLADDER   SPHINCTER (−) of ODDI

CHOLECYSTOKININ → Venous blood from duodenal mucosa → General Circulation → Arterial blood to stimulate muscular wall of Gall Bladder

[Cholecystokinin is probably the main factor controlling Gall Bladder
contraction in man]

SMALL INTESTINE

The Small Intestine is a long muscular tube - over 6 metres in length. It receives CHYME in small quantities from the STOMACH; PANCREATIC JUICE from the PANCREAS; BILE from the GALL BLADDER.

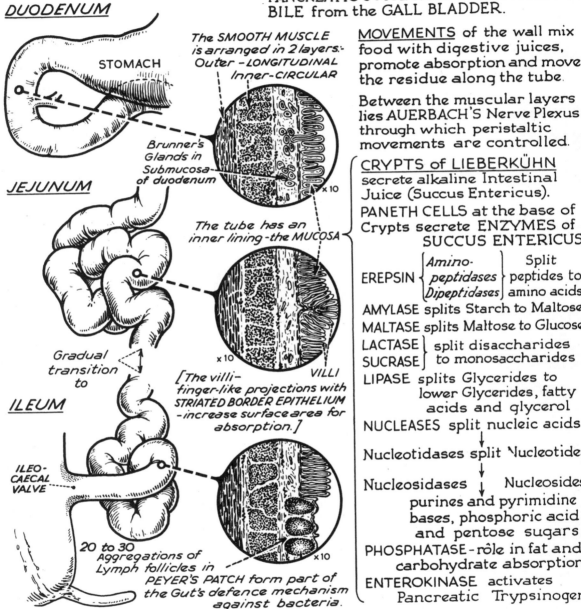

DUODENUM

STOMACH

The SMOOTH MUSCLE is arranged in 2 layers:- Outer - LONGITUDINAL Inner - CIRCULAR

Brunner's Glands in Submucosa of duodenum

×10

JEJUNUM

The tube has an inner lining - the MUCOSA

×10

VILLI

Gradual transition to

[The villi - finger-like projections with STRIATED BORDER EPITHELIUM - increase surface area for absorption.]

ILEUM

ILEO-CAECAL VALVE

20 to 30 Aggregations of Lymph follicles in PEYER'S PATCH form part of the Gut's defence mechanism against bacteria.

×10

MOVEMENTS of the wall mix food with digestive juices, promote absorption and move the residue along the tube.

Between the muscular layers lies AUERBACH'S Nerve Plexus through which peristaltic movements are controlled.

CRYPTS of LIEBERKÜHN secrete alkaline Intestinal Juice (Succus Entericus).

PANETH CELLS at the base of Crypts secrete ENZYMES of SUCCUS ENTERICUS.

EREPSIN { Amino-peptidases, Dipeptidases } Split peptides to amino acids

AMYLASE splits Starch to Maltose

MALTASE splits Maltose to Glucose

LACTASE, SUCRASE } split disaccharides to monosaccharides

LIPASE splits Glycerides to lower Glycerides, fatty acids and glycerol

NUCLEASES split nucleic acids

↓

Nucleotidases split Nucleotides

↓

Nucleosidases ↓ Nucleosides purines and pyrimidine bases, phosphoric acid and pentose sugars

PHOSPHATASE - rôle in fat and carbohydrate absorption

ENTEROKINASE activates Pancreatic Trypsinogen.

The BASIC PATTERN of the GUT WALL

The wall of the digestive tube shows a basic structural pattern.
FOUR coats are seen in transverse section.

STRUCTURE | FUNCTIONS

1. MUCOUS COAT *(or MUCOSA)* ———

Layer in close contact with gut contents: specialized for –
SECRETION
ABSORPTION – nutrients and hormones to blood stream.

SURFACE EPITHELIUM *[type varies with site and function]* with its GLANDS

LOOSE FIBROUS TISSUE *(lamina propria)* with capillaries and lymphatic vessels.

MUSCULARIS MUCOSAE *(thin layers of smooth muscle)*

LYMPHATIC TISSUE

MOBILITY (can continually change degree of folding and the surface of contact with gut contents).

DEFENCE against bacteria.

2. SUBMUCOUS COAT *(or SUBMUCOSA)*

DENSE FIBROUS TISSUE in which lie bloodvessels, lymphatic vessels and MEISSNER'S NERVE PLEXUS [Glands in oesophagus and 1st part of duodenum] LYMPHATIC TISSUE

STRONG LAYER of SUPPLY to specialized mucosa (Bloodvessels supply needs and remove absorbed materials).

CO-ORDINATION of motor and secretory activities of mucosa.

3. MUSCULAR COAT *(or MUSCULARIS EXTERNA)*

SMOOTH MUSCLE LAYERS
Inner–*Circular arrangement*
Outer – *Longitudinal arrangement*
Between them, blood- and lymphatic vessels and AUERBACH'S NERVE PLEXUS [Some striated muscle in oesophagus]

MOVEMENT

Controls diameter of tube.
Mixes contents.
Propagates contents along tube.

Nervous elements co-ordinate secretory and muscular activities.

4. SEROUS COAT *(or SEROSA)*

FIBROUS TISSUE with fat, blood- and lymphatic vessels
MESOTHELIUM
Where tube is suspended by a MESENTERY the serosa is formed by visceral layer of Peritoneum

[Where there is no mesentery – replaced by fibrous tissue which merges with surrounding fibrous tissue]

Carries nerves, blood- and lymphatic vessels to and from mesentery.

Forms smooth, moist membrane which reduces friction between contacting surfaces in the peritoneal cavity.

(After GARVEN)

67

INTESTINAL JUICE

Two types of secretion are formed in the intestine.

1. ALKALINE secretion (with Mucus) secreted by Submucosal Brünner's Glands in first part of Duodenum to protect the mucosa against ACID of Stomach.   ACID CHYME → DUODENAL MUCOSA → Hormone - DUOCRININ ⟶ BLOOD STREAM → stimulates Brünner's Glands to secrete.

2. SUCCUS ENTERICUS secreted by Crypts of Lieberkühn.
   Its secretion is probably under two types of control :- NERVOUS and HUMORAL

<u>NERVOUS</u>   Stimulation of Parasympathetic Nerves ⟶

A Local Reflex

STIMULUS	*INGOING PATHWAY*	*CELLS IN AUERBACH'S PLEXUS*	*OUTGOING PATHWAY*	*EFFECT*
Food distending intestinal wall →	Intrinsic nerves of the plexus in the wall of the small intestine	*Secretion* *Pressure of Bolus on Gut Wall*	Intrinsic nerves of the plexus in the wall of the small intestine	Secretion of Succus Entericus by Crypt Cells

LUNGS

HEART

LIVER   STOMACH

<u>HUMORAL</u>

Acid CHYME stimulates duodenal mucosa to secrete
↓

BOLUS

ENTEROCRININ → Blood stream → General Circulation → { Arterial blood to Mucosa of Small Intestine stimulates intestinal glands → Secretion of Succus Entericus

SECRETIN ⟶ Blood stream → General Circulation →

MOVEMENTS of the SMALL INTESTINE

The Duodenum receives food in small quantities from the Stomach. The mixture of food and digestive juices — Chyme — is passed along the length of the Small Intestine.

TWO TYPES of MOVEMENT

SEGMENTATION —
Rhythmical alternating contractions and relaxations

— These "shuttling" movements serve to mix CHYME and to bring it into contact with the absorptive mucosa, i.e. DIGESTION and ABSORPTION are promoted.

This type of movement is MYOGENIC, i.e. it is the property of the smooth muscle cells. It does not depend on a nervous mechanism.

PERISTALSIS —
Food probably acts as stimulus to stretch receptors in muscle and perhaps in mucous membrane

Circular muscle behind bolus CONTRACTS   Muscle in front of bolus RELAXES

BOLUS

— Waves of this contraction move the food along the canal.

The contraction behind the bolus sweeps it into the relaxed portion of the tube ahead.

This type of movement is NEUROGENIC, i.e. it is carried out through a "Local" Reflex mediated through INTRINSIC nerve plexuses within the wall of the tube.

EMPTYING

ILEO-CAECAL VALVE ------- Formed by thickening of circular muscle layer

ILEUM

CAECUM

opens and closes during digestion to allow spurts of fluid material from the Ileum to enter the Large Intestine.

This is a "Central" Reflex initiated when food enters the stomach and carried out through EXTRINSIC nerves to wall of tube.

Meals of different composition travel along the intestine at different rates. DIGESTION and ABSORPTION of food are usually complete when the Residue reaches the Ileo-caecal valve.

ABSORPTION in SMALL INTESTINE

Absorption of most digested foodstuffs occurs in the Small Intestine through the Striated Border Epithelium covering the Villi.

Amino Acids; Sugars; Minerals; Glycerol; some Fatty Acids and Vitamins } into CAPILLARIES

Glycerides and some Fatty Acids and Fat-soluble Vitamins } into LACTEALS

Absorption is aided by movements of villi ("Villus Pump" mechanism) brought about by contraction of smooth muscle – extensions of muscularis mucosae – present in core of villus and attached to lacteal and to basement membrane of epithelium.

Movements of the villi are regulated through NERVE PLEXUS of MEISSNER and perhaps stimulated by VILLIKININ – a hormone liberated to Blood Stream by Duodenal mucosa.

Movements of the muscular coats are regulated through the NERVE PLEXUS of AUERBACH.

The Intestines are suspended from the Posterior Abdominal wall by a delicate membrane –the MESENTERY–which carries the mesenteric arteries and veins to and from the gut; and lymphatic vessels on their way via the mesenteric lymph nodes to Thoracic Duct.

[Structures shown are present in each villus]

------Mesentery

Branches of Mesenteric Vein

Branches of Mesenteric Artery

Lymphatic Vessels

Nerves

Absorption is not just a process of simple diffusion of substances from areas of high to areas of low concentration. Movement of ions can take place against a concentration gradient. In other words — Absorption is often an active process involving the use of energy by the epithelial cells.

TRANSPORT of ABSORBED FOODSTUFFS

After absorption the NUTRIENTS are transported in:—

(a) **BLOOD** through
   Mesenteric Veins
        to
   Portal Vein
        to
   LIVER— *Many substances undergo further changes; some are stored; some are passed on via↓*

Hepatic Veins → Inferior
   Vena Cava →
   Heart and
   Systemic
   Circulation

(b) **LYMPH** in
   Lymphatic Vessels
        to
   THORACIC DUCT
   and via a large vein
   in the neck to
   Systemic Circulation
        ↓
   which then distributes
   Food to ALL TISSUES
   of the BODY

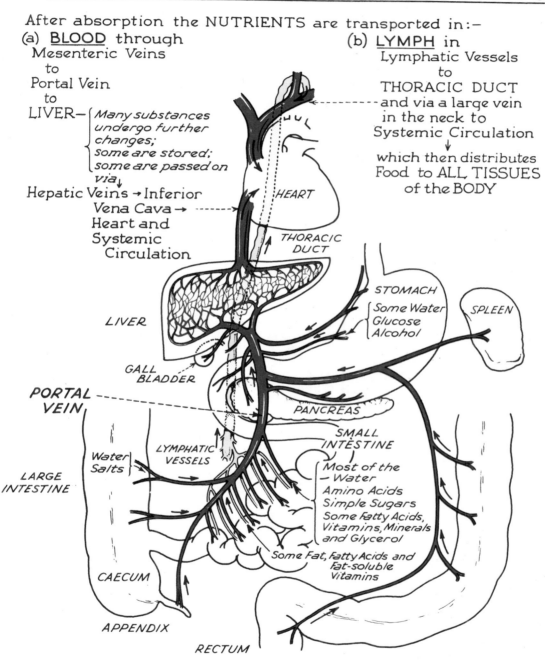

HEART

THORACIC DUCT

LIVER

STOMACH
*Some Water
Glucose
Alcohol*

SPLEEN

GALL BLADDER

PORTAL VEIN

PANCREAS

SMALL INTESTINE
*Most of the
Water
Amino Acids
Simple Sugars
Some Fatty Acids,
Vitamins, Minerals
and Glycerol*

*Some Fat, Fatty Acids and
Fat-soluble
Vitamins*

Water
Salts

LYMPHATIC VESSELS

LARGE INTESTINE

CAECUM

APPENDIX

RECTUM

LARGE INTESTINE

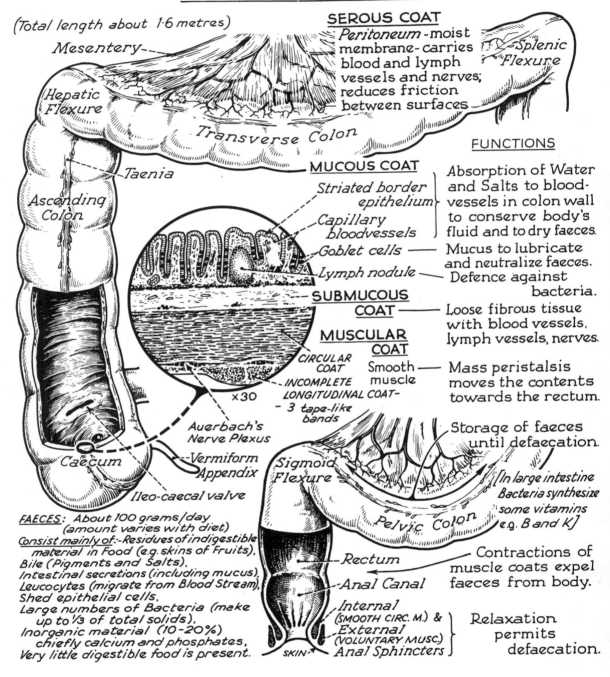

(Total length about 1·6 metres)

Mesentery--

Hepatic
Flexure

Ascending
Colon

--Taenia

Transverse Colon

--Splenic
Flexure

SEROUS COAT
Peritoneum -moist
membrane- carries
blood and lymph
vessels and nerves;
reduces friction
between surfaces

MUCOUS COAT
Striated border
epithelium
Capillary
bloodvessels
Goblet cells ———
Lymph nodule ——

SUBMUCOUS
COAT ———

MUSCULAR
COAT
CIRCULAR
COAT
Smooth
muscle
INCOMPLETE
LONGITUDINAL COAT-
- 3 tape-like
bands

×30

Auerbach's
Nerve Plexus

Caecum

--Vermiform
Appendix

Ileo-caecal valve

Sigmoid
Flexure

Pelvic Colon

-Rectum

-Anal Canal

Internal
(SMOOTH CIRC. M.) &
External
(VOLUNTARY MUSC.)
Anal Sphincters

SKIN

FUNCTIONS

Absorption of Water
and Salts to blood-
vessels in colon wall
to conserve body's
fluid and to dry faeces.

Mucus to lubricate
and neutralize faeces.

Defence against
bacteria.

Loose fibrous tissue
with blood vessels,
lymph vessels, nerves.

Mass peristalsis
moves the contents
towards the rectum.

Storage of faeces
until defaecation.

[In large intestine
Bacteria synthesize
some vitamins
e.g. B and K]

Contractions of
muscle coats expel
faeces from body.

Relaxation
permits
defaecation.

FAECES: About 100 grams/day
(amount varies with diet)
Consist mainly of:-Residues of indigestible
material in Food (e.g. skins of Fruits),
Bile (Pigments and Salts).
Intestinal secretions (including mucus),
Leucocytes (migrate from Blood Stream),
Shed epithelial cells.
Large numbers of Bacteria (make
up to ⅓ of total solids).
Inorganic material (10-20%)
chiefly calcium and phosphates,
Very little digestible food is present.

72

MOVEMENTS of the LARGE INTESTINE

The Ileo-caecal Valve opens and closes during digestion. Peristaltic waves sweep semi-fluid contents of Ileum through the relaxed valve. During its stay in the Large Intestine faecal matter is subjected to –

TWO TYPES OF MOVEMENT

"TONE WAVES" run forwards – and backwards from Cannon's Point to the Ileum – (sometimes mistakenly called "Anti-peristalsis") These waves "die out"

→ Ensure prolonged contact of contents with mucosa —and promote absorption of Water and Salt from faeces —

This type of movement is MYOGENIC

MASS PERISTALSIS Strong waves at infrequent intervals start at upper end of Ascending Colon

→ Empty Transverse Colon and sweep faeces into the Descending and Pelvic Colons and into Rectum – —*NEUROGENIC*

[Reflex often initiated by passage of food into Stomach (Gastro-colic Reflex)]

EMPTYING

(DEFAECATION)

Complex **REFLEX** *act*

Stimulus Passage of faeces into the Rectum distends wall

[+ passage through anal canal]

Sensations rise to level of consciousness → Voluntary decision → Impulses to inhibit or permit reflex evacuation.

Outgoing nerve messages → Powerful peristaltic contractions of Descending Colon, Pelvic Colon and Rectum.

[*Preceded by inspiratory descent of diaphragm and voluntary contraction of abdominal muscles to raise intra-abdominal pressure*]

Contraction of Pelvic Floor muscles with Relaxation of Anal Sphincter

↓

Evacuation of Faeces

INNERVATION of the GUT WALL

The movements of the Alimentary Canal are carried out automatically and, for the most part, beneath the level of consciousness.

Like other "automatic" systems in the body the gut wall is supplied by nerves of the AUTONOMIC NERVOUS SYSTEM. This has TWO separate parts – SYMPATHETIC and PARASYMPATHETIC. They work together to achieve <u>balanced control</u> of events.

PRE-GANGLIONIC FIBRE

COLLATERAL GANGLION

POST-GANGLIONIC FIBRE

SYMPATHETIC Nerves when active appear to release at their synapses a chemical substance, *ACETYLCHOLINE*, and at their endings in the muscle, *NORADRENALINE*. This *inhibits* (−) [i.e. causes relaxation of] the gut wall, and *stimulates* (+) [i.e. causes contraction or closure of] the sphincters.

LONGITUDINAL MUSCLE COAT
AUERBACH'S PLEXUS
CIRCULAR MUSCLE COAT
MUCOSA AND SUBMUCOSA
LUMEN OF GUT
SPHINCTER
AUERBACH'S PLEXUS

PRE-GANGLIONIC FIBRE

PARASYMPATHETIC Nerves when active release *ACETYLCHOLINE* at their synapses and at their endings in the muscle. This *stimulates* (+) the gut wall to contract and *inhibits* (−), or relaxes, the sphincters.

74

NERVOUS CONTROL of GUT MOVEMENTS

Movements in the wall of the Alimentary Canal are

either < (a) *MYOGENIC* — *a property of the smooth muscle.*
       < (b) *NEUROGENIC* – *dependent on the Intrinsic Nerve Plexuses.*

They can occur even after *Extrinsic* nerves to the tract have been cut. Normally, however, impulses travelling in these nerves of the SYMPATHETIC and PARASYMPATHETIC Systems, from the CONTROLLING CENTRES of the AUTONOMIC NERVOUS SYSTEM in the BRAIN, *influence* and *co-ordinate* events in the whole tract.

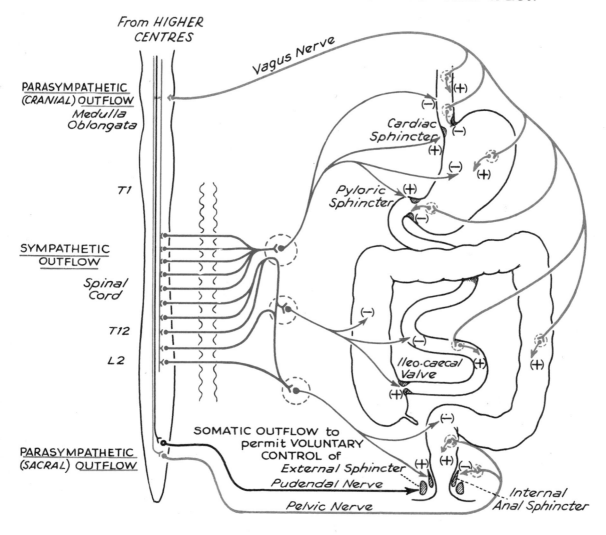

From HIGHER CENTRES

Vagus Nerve

PARASYMPATHETIC (CRANIAL) OUTFLOW
Medulla Oblongata

Cardiac Sphincter

Pyloric Sphincter

T1

SYMPATHETIC OUTFLOW

Spinal Cord

T12

L2

Ileo-caecal Valve

SOMATIC OUTFLOW to permit VOLUNTARY CONTROL of
External Sphincter
Pudendal Nerve
Pelvic Nerve

PARASYMPATHETIC (SACRAL) OUTFLOW

Internal Anal Sphincter

CHAPTER 4

TRANSPORT SYSTEM

The HEART, BLOOD VESSELS and BODY FLUIDS: HAEMOPOIETIC SYSTEM

CARDIOVASCULAR SYSTEM

The CIRCULATORY System

Chief TRANSPORT System of the body

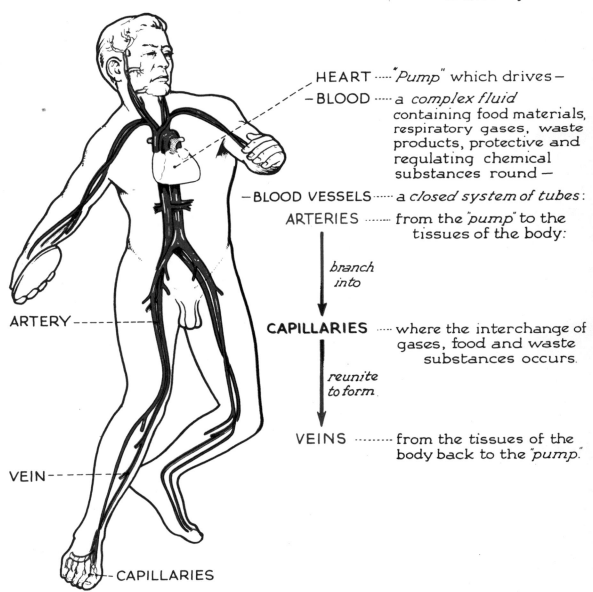

HEART ----- "*Pump*" which drives —

—BLOOD ----- a *complex fluid* containing food materials, respiratory gases, waste products, protective and regulating chemical substances round —

—BLOOD VESSELS ----- a *closed system of tubes*:

ARTERIES ----- from the "*pump*" to the tissues of the body:

branch into

CAPILLARIES ----- where the interchange of gases, food and waste substances occurs.

reunite to form

VEINS ----- from the tissues of the body back to the "*pump*."

ARTERY

VEIN

CAPILLARIES

GENERAL COURSE of the CIRCULATION

CAPILLARY BED – HEAD, NECK, UPPER LIMBS

PULMONARY CIRCULATION

To PULMONARY (or Lesser) CIRCULATION

R.A.

L.A.

L.V.

R.V.

To SYSTEMIC (or Greater) CIRCULATION

HEPATIC CIRCULATION

PORTAL VEIN

CAPILLARY BED – SPLEEN, DIGESTIVE TRACT

RENAL CIRCULATION

CAPILLARY BED – TRUNK, LOWER LIMBS

HEART

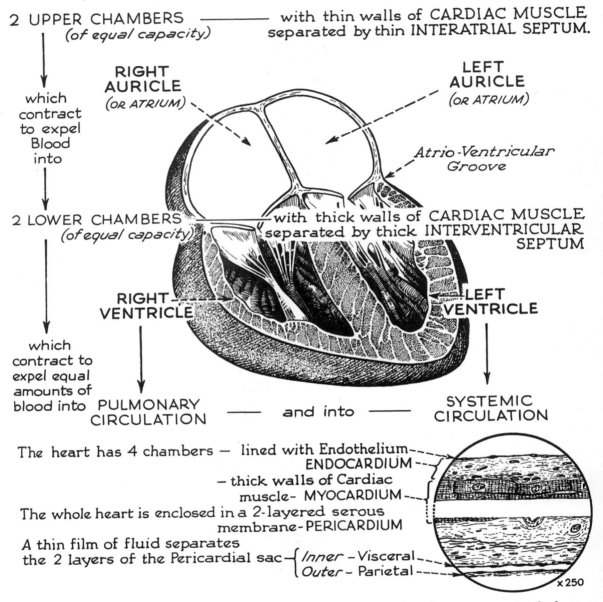

2 UPPER CHAMBERS ——————— with thin walls of CARDIAC MUSCLE
(of equal capacity) separated by thin INTERATRIAL SEPTUM.

RIGHT AURICLE
(OR ATRIUM)

LEFT AURICLE
(OR ATRIUM)

Atrio-Ventricular Groove

which contract to expel Blood into

2 LOWER CHAMBERS ———— with thick walls of CARDIAC MUSCLE
(of equal capacity) separated by thick INTERVENTRICULAR SEPTUM

RIGHT VENTRICLE

LEFT VENTRICLE

which contract to expel equal amounts of blood into

PULMONARY CIRCULATION —— and into —— SYSTEMIC CIRCULATION

The heart has 4 chambers — lined with Endothelium—ENDOCARDIUM

— thick walls of Cardiac muscle- MYOCARDIUM

The whole heart is enclosed in a 2-layered serous membrane-PERICARDIUM

A thin film of fluid separates the 2 layers of the Pericardial sac {*Inner – Visceral* / *Outer – Parietal*}

× 250

The Cardiac muscle of the Auricles is completely separated from the Cardiac muscle of the Ventricles by a ring of Fibrous Tissue — at the AURICULO -VENTRICULAR (ATRIO-VENTRICULAR) Groove.

HEART

This is a *diagrammatic* section through the heart.

<u>HEART VALVES</u> have a core of Fibrous Tissue ----------
covered on both sides with *Endothelium*

Extensions from
<u>AURICULO-VENTRICULAR (A-V)
FIBROUS RING</u>

Designed to
allow blood
to flow in
one direction
only —
from AURICLE
to VENTRICLE—
and on into
ARTERIES

The A-V valves
are attached by
thin CHORDAE
    TENDINEAE
        to
extensions of
CARDIAC MUSCLE —
PAPILLARY MUSCLES

These contract
when ventricles contract
and pull on Chordae
Tendineae so that valve
flaps cannot be everted.

TO
BODY
TISSUES

AORTA

P.A.

RIGHT
LUNG

PULMONARY ART.

LEFT
LUNG

PULM.
VEINS

PULM.
VEINS

S.V.C.

I.V.C.

SEMILUNAR
    VALVES
*(each with three
flaps)* prevent
BACKFLOW
from
PULMONARY
ARTERY and
AORTA.

TRICUSPID
VALVE

MITRAL
VALVE

*The Great Veins do not have valves guarding their entrance to the heart.
Thickening and contraction of the muscle around their mouths
                prevent BACKFLOW of blood from heart.*

HEART

The human heart is really a *DOUBLE PUMP*.

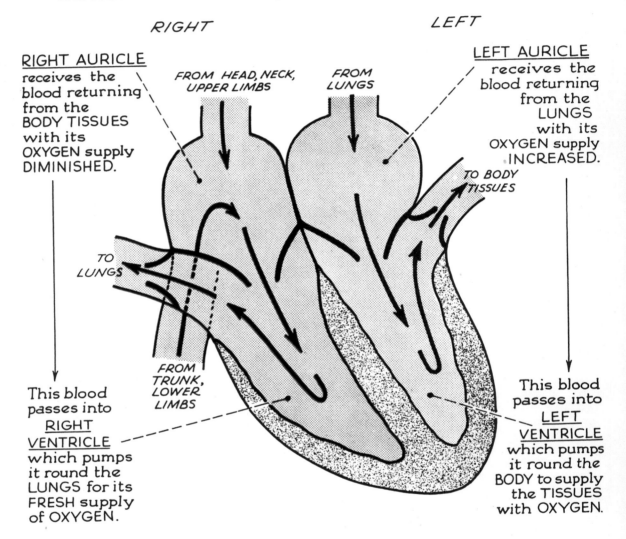

RIGHT   LEFT

RIGHT AURICLE
receives the
blood returning
from the
BODY TISSUES
with its
OXYGEN supply
DIMINISHED.

FROM HEAD, NECK,
UPPER LIMBS

FROM
LUNGS

LEFT AURICLE
receives the
blood returning
from the
LUNGS
with its
OXYGEN supply
INCREASED.

TO BODY
TISSUES

TO
LUNGS

FROM
TRUNK,
LOWER
LIMBS

This blood
passes into
RIGHT
VENTRICLE
which pumps
it round the
LUNGS for its
FRESH supply
of OXYGEN.

This blood
passes into
LEFT
VENTRICLE
which pumps
it round the
BODY to supply
the TISSUES
with OXYGEN.

*This diagram simplifies the structure of the heart to make it
easier to understand the function of its various parts.*

CARDIAC CYCLE

Diagrammatic representation of the sequence of events in the heart during ONE heart beat.

DIASTOLE *[Period of Relaxation]* ⟶

Auricles filling — *Auricles filling Pressure rising* — *Auricles full* — *Flow of blood slows and ceases*

All valves closed — *A-V valves pushed open* — *A-V valves open*

Ventricles filling — *Ventricles full*

ISOMETRIC RELAXATION — PERIOD OF RAPID FILLING — PERIOD OF DIASTASIS

SYSTOLE
[Period of Contraction]

Auricles contract

Ventricles continue to contract. Pressure continues to rise→ Semilunar valves pushed open. Blood ejected into Pulm. A. and Aorta. — *Ventricles now contract. Pressure rising (closes A-V valves)* — *Ventricles still relaxed now overfilled and stretched.*

PERIOD OF EJECTION — ISOMETRIC CONTRACTION — AURICULAR SYSTOLE

VENTRICULAR SYSTOLE

The total cycle of events takes about 0·8 second when heart is beating 75 times per minute.

83

HEART SOUNDS

During each CARDIAC CYCLE 2 HEART SOUNDS can be heard through a STETHOSCOPE applied to the CHEST WALL.

1st HEART SOUND

is chiefly due to the coming together of the ATRIO-VENTRICULAR VALVE FLAPS at the beginning of VENTRICULAR SYSTOLE

AORTIC VALVE

PULMONARY VALVE

TRICUSPID VALVE

MITRAL VALVE

()'s indicate areas at which sounds produced by each valve are best heard

2nd HEART SOUND

is chiefly due to the closure of the SEMILUNAR VALVES at the beginning of VENTRICULAR DIASTOLE

Lower Pitch
Softer
Longer

Higher Pitch
Sharper
Shorter

The sounds may be represented phonetically:-

LUBB | DUP | LUBB | DUP

| 1 | 2 | 1 | 2 | 1 | 2 |

VENTRICULAR SYSTOLE — VENTRICULAR DIASTOLE

They are repeated with every CARDIAC CYCLE i.e. about 70 times per minute in the average healthy adult.

If the valves have been damaged by disease additional murmurs can be heard as the blood flows forwards through narrowed valves or leaks backwards through incompetent valves.

Occasionally additional sounds are heard over normal healthy hearts.

ORIGIN and CONDUCTION of the HEART BEAT

The rhythmic contraction of the heart is called the HEART BEAT. The impulse to contract is generated in specialized NODAL TISSUE in the wall of the RIGHT AURICLE.

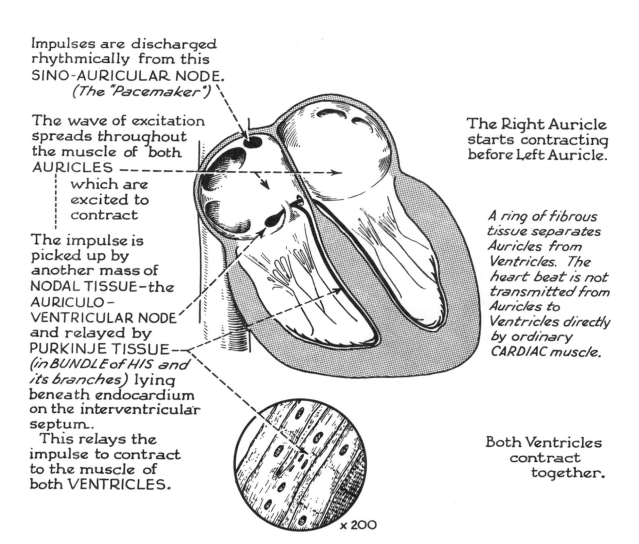

Impulses are discharged rhythmically from this SINO-AURICULAR NODE. *(The "Pacemaker")*

The wave of excitation spreads throughout the muscle of both AURICLES — — — — — which are excited to contract

The impulse is picked up by another mass of NODAL TISSUE—the AURICULO-VENTRICULAR NODE and relayed by PURKINJE TISSUE—*(in BUNDLE of HIS and its branches)* lying beneath endocardium on the interventricular septum.

This relays the impulse to contract to the muscle of both VENTRICLES.

The Right Auricle starts contracting before Left Auricle.

A ring of fibrous tissue separates Auricles from Ventricles. The heart beat is not transmitted from Auricles to Ventricles directly by ordinary CARDIAC muscle.

x 200

Both Ventricles contract together.

The wave of excitation is accompanied by an electrical change which is followed within 0·02 seconds by contraction of the cardiac muscle.

ELECTROCARDIOGRAM

The wave of excitation spreading through the heart wall is accompanied by electrical changes. *(Like Nerve and Skeletal Muscle, active Cardiac Muscle is electrically negative relative to resting Cardiac Muscle ahead of the zone of excitation.)* The electrical currents produced are conducted to the surface of the body and can be picked up, amplified and recorded by a special instrument – the <u>ELECTROCARDIOGRAPH</u>.

The record obtained is the *ELECTROCARDIOGRAM : (E.C.G.)*

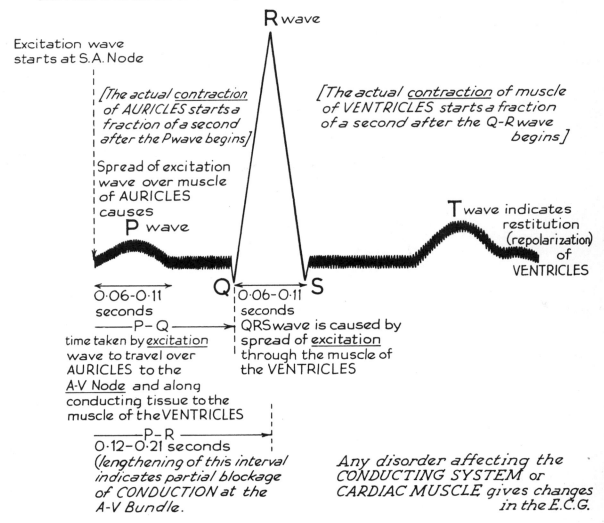

R wave

Excitation wave starts at S.A. Node

[The actual <u>contraction</u> of AURICLES starts a fraction of a second after the P wave begins]

[The actual <u>contraction</u> of muscle of VENTRICLES starts a fraction of a second after the Q-R wave begins]

Spread of excitation wave over muscle of AURICLES causes P wave

T wave indicates restitution (repolarization) of VENTRICLES

Q   S

0·06-0·11 seconds
———P–Q———
time taken by <u>excitation</u> wave to travel over AURICLES to the <u>A-V Node</u> and along conducting tissue to the muscle of the VENTRICLES
———P–R———
0·12–0·21 seconds
(lengthening of this interval indicates partial blockage of CONDUCTION at the A-V Bundle.

0·06-0·11 seconds
QRS wave is caused by spread of <u>excitation</u> through the muscle of the VENTRICLES

Any disorder affecting the CONDUCTING SYSTEM or CARDIAC MUSCLE gives changes in the E.C.G.

NERVOUS REGULATION of ACTION of HEART

Although the heart initiates its own impulse to contraction its activity is finely adjusted to meet the body's constantly changing needs by nervous impulses discharged from CONTROLLING CENTRES in the BRAIN and SPINAL CORD along PARASYMPATHETIC and SYMPATHETIC OUTFLOWS.

ACTION of PARASYMPATHETIC —

Continuous stream of impulses →
tends to restrain (—) Heart's Action:-
RATE of CONTRACTION DECREASED,
CONDUCTIVITY DEPRESSED,
FORCE of CONTRACTION DECREASED,
EXCITABILITY DECREASED,
Refractory Period shortened.
PARASYMPATHETIC OUTFLOW

[Variation in this 'VAGAL TONE' is chief factor in giving alteration of Heart Rate]

Highest Centres are probably in CEREBRAL CORTEX : Psychic and emotional factors can influence heart's action.

MIDDLE GROUP OF NEURONES

CARDIO-INHIBITOR CENTRE

VAGUS NUCLEUS

HYPOTHALAMUS

POSTERIOR GROUP OF NEURONES send impulses to MEDULLA OBLONGATA CARDIO-ACCELERATOR CENTRE

RIGHT VAGUS · LEFT VAGUS

SYMPATHETIC OUTFLOW

ACCELERATOR NERVES

T1

Post-ganglionic fibres of both Parasympathetic and Sympathetic form a CARDIAC PLEXUS of nerves before final distribution to the heart muscle.

ACTION of SYMPATHETIC —

Constantly transmits streams of impulses →
tend to accelerate (+) Heart's Action:-
RATE of CONTRACTION INCREASED,
CONDUCTIVITY INCREASED,
FORCE of CONTRACTION INCREASED,
EXCITABILITY INCREASED,
Refractory Period lengthened.

The Parasympathetic and Sympathetic have opposite effects on heart's action. Both are constantly in action but finely balanced to meet the body's needs.

Stimulation of Symp.
Inhibition of Para. } Acceleration } of
Stimulation of Para. } Inhibition } Heart's
Inhibition of Symp. } Action

CARDIAC REFLEXES

There are INGOING *(SENSORY)* fibres probably in both PARASYMPATHETIC and SYMPATHETIC nerves which convey information to the CENTRES about events taking place in the heart.

These afferent messages do not normally reach consciousness. They are important as the AFFERENT pathways in CARDIAC REFLEXES by means of which heart's action is adjusted to body's needs.

CARDIO-ACCELERATOR REFLEXES

AUTONOMIC CENTRES IN
MEDULLA
OBLONGATA

BAINBRIDGE REFLEX

AFFERENTS
Cranial X (Vagus)

RECEPTORS
Stretch receptors near mouths of GREAT VEINS stimulated

Stretching of Receptors in Great Veins by rise in pressure in Right Auricle

STIMULUS
Increased return of venous blood to Right Auricle

EFFERENTS
Decreased number of impulses sent out along PARASYMPATHETIC OUTFLOW

Increased number of impulses sent out along SYMPATHETIC OUTFLOW
to HEART MUSCLE
↓
ACCELERATION
of
HEART'S ACTION

This has been regarded for almost 50 years as an important adaptive mechanism whereby heart rate and force of contraction were reflexly adjusted to quantity of venous blood returning to the heart. Recently some doubt has been cast on the exact rôle of this reflex in the normal regulation of the heart rate.

CARDIAC REFLEXES

CARDIO-INHIBITORY REFLEXES

AUTONOMIC
CENTRES IN
MEDULLA
OBLONGATA

AFFERENTS

["Buffer Nerves"]

IX (GLOSSOPHARYNGEAL)

INTERNAL
CAROTID
ARTERY

X (VAGUS)

RECEPTORS
Stretch (or Baro-)
receptors in CAROTID
SINUS

ARCH of
AORTA
and probably
in MYOCARDIUM

Stretching of Aorta
by increased blood
pressure

Pressure
high in
Left
Ventricle

EFFERENTS

Stimulation
of
PARASYMPATHETIC
OUTFLOW

and

Inhibition
of
SYMPATHETIC
OUTFLOW

STIMULUS — —→ Increased arterial
blood pressure

Stimulation of other afferents e.g. from SPLANCHNIC
region of abdominal cavity and overstimulation of
any PAIN-sensitive nerves may also give reflex ——→

EFFECT

SLOWING of HEART'S
ACTION
and Fall in Blood
Pressure

This reflex is constantly in action —
Important Adaptive Mechanism — As Arterial Blood Pressure is raised
by ejection of blood from Ventricle in Systole this reflex is brought
into play to adjust force and frequency with which heart is ejecting
blood into the arterial system ——→ prevent undue rise in blood pressure.

89

CARDIAC OUTPUT

The Volume of Blood expelled from the heart can be measured as follows :-

The AMOUNT of OXYGEN taken up by the lungs per minute is measured by a SPIROMETER.

A semi-rigid tube is inserted into a VEIN in the arm and passed along into the RIGHT AURICLE of the heart — to obtain a sample of MIXED VENOUS BLOOD — which has given up some of its oxygen to the tissues.
The OXYGEN content is analysed.

<u>250 ml OXYGEN</u> are removed from the lungs by blood each minute.

A needle is inserted into an ARTERY in the leg and a sample of ARTERIAL BLOOD — which has received its fresh oxygen supply in the lungs — is obtained.
The OXYGEN content is analysed.

<u>100 ml VENOUS Blood hold 14 ml OXYGEN</u>.

(After WISHART)

Each 100 ml blood gains 5 ml OXYGEN *as it passes through the lungs. The blood in the lungs takes up* 250 ml OXYGEN *from the atmosphere per minute.*

<u>100 ml ARTERIAL Blood hold 19 ml OXYGEN</u>.

Therefore there must be $\left(\frac{250}{5} \times 100\right)$ ml *[i.e. 5,000 ml] of blood leaving the Right Ventricle and passing through the Lungs to the Left Atrium per minute to take up this* 250 ml OXYGEN.

The same volume of blood must leave the Left Ventricle and enter the Aorta in the same time otherwise blood would soon be dammed back in the lungs. i.e. If heart contracts 72 times per minute STROKE VOLUME $= \frac{5000}{72} \doteqdot$ 70 ml *per beat for each Ventricle.*

Cardiac Output can be increased many times *(up to 30 litres per min)* in Exercise —
   —partly by increase in Heart Rate; partly by increase in Stroke Volume.

BLOOD VESSELS

The system of tubes through which the heart pumps blood.

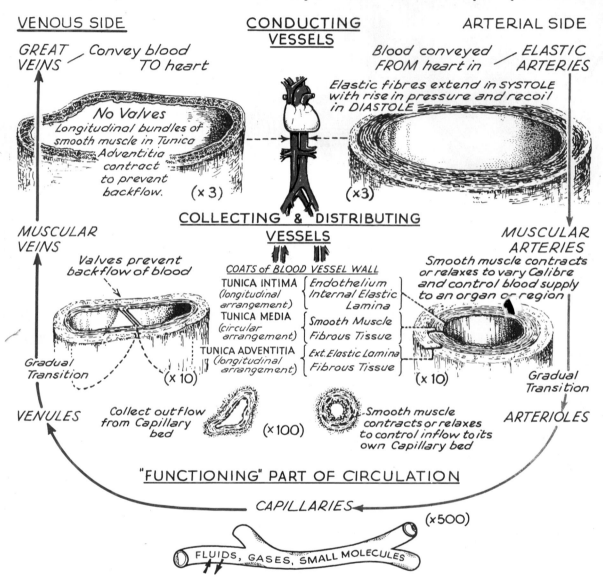

VENOUS SIDE

GREAT VEINS — Convey blood TO heart

No Valves
Longitudinal bundles of smooth muscle in Tunica Adventitia contract to prevent backflow. (x 3)

CONDUCTING VESSELS

Blood conveyed FROM heart in — *ELASTIC ARTERIES*

Elastic fibres extend in SYSTOLE with rise in pressure and recoil in DIASTOLE (x3)

ARTERIAL SIDE

MUSCULAR VEINS

Valves prevent backflow of blood

Gradual Transition (x 10)

COLLECTING & DISTRIBUTING VESSELS

COATS of BLOOD VESSEL WALL

TUNICA INTIMA (*longitudinal arrangement*) { Endothelium / Internal Elastic Lamina

TUNICA MEDIA (*circular arrangement*) { Smooth Muscle / Fibrous Tissue

TUNICA ADVENTITIA (*longitudinal arrangement*) { Ext. Elastic Lamina / Fibrous Tissue

MUSCULAR ARTERIES
Smooth muscle contracts or relaxes to vary Calibre and control blood supply to an organ or region (x 10)

Gradual Transition

VENULES

Collect outflow from Capillary bed (x100)

Smooth muscle contracts or relaxes to control inflow to its own Capillary bed

ARTERIOLES

"FUNCTIONING" PART OF CIRCULATION

CAPILLARIES (x500)

FLUIDS, GASES, SMALL MOLECULES

Only from CAPILLARIES can Blood give up food and oxygen to tissues; and receive waste products and carbon dioxide from tissues.

BLOOD PRESSURE

Each Ventricle at each heart beat ejects forcibly about 70c.c. of Blood into the Blood Vessels. All of this blood cannot pass through arterioles into capillaries and veins during the hearts contraction. This means that roughly 5/8 of the <u>CARDIAC OUTPUT</u> at each heart beat has to be stored during systole and passed on during diastole.

    <u>CONDUCTING ARTERIES</u> are always more or less stretched.

    <u>PERIPHERAL RESISTANCE</u> is offered to the passage of blood from arterial to venous side of the system chiefly by partial constriction ("Tone") of smooth muscle in walls of Arterioles. (*The calibre is regulated by action of Parasympathetic and Sympathetic Nervous System* — [see pages 95-97]

    These factors are largely responsible for the considerable pressure of the blood in the Arterial System. The pressure is highest at the height of the heart's contraction, i.e. <u>SYSTOLIC BLOOD PRESSURE</u>, and lowest when the heart is relaxing, i.e. <u>DIASTOLIC BLOOD PRESSURE</u>.

The pressure is LOWEST as blood drains into RIGHT AURICLE at end of DIASTOLE.

The pressure is HIGHEST as blood leaves the LEFT VENTRICLE at end of SYSTOLE.

GREAT VEINS — *near heart about 4mm Hg*

R.A.

L.V.

ARTERIAL SYSTEM

AORTA — Systolic Blood Pressure = 120 mm Hg

Pressure is always lower in *DIASTOLE* e.g 80 mm Hg

It falls off gradually

MUSCULAR VEINS 12 mm Hg

Larger MUSCULAR ARTERIES 110 mm Hg

VENULES 16mm Hg

ARTERIOLES 40 mm Hg

Most of the Energy imparted to Blood by Heart's Contraction has been spent by the time the Blood reaches the venous side of circulation.

CAPILLARIES 16mm Hg ← 30 mm Hg

NOTE :- *Any alteration in the* <u>TOTAL AMOUNT</u> *or* <u>VISCOSITY</u> *of Blood will also affect BLOOD PRESSURE.*

MEASUREMENT of ARTERIAL BLOOD PRESSURE

The Arterial Blood Pressure is measured in man by means of a *SPHYGMOMANOMETER*.

This consists of a RUBBER BAG (*covered with a cloth envelope*) which is wrapped round the UPPER ARM over the BRACHIAL ARTERY.

One tube connects the inside of the bag with a MANOMETER containing MERCURY.

Another tube connects the inside of the bag to a hand operated PUMP with a release VALVE.

Mercury Manometer

Cross-section of Arm

Cuff

Release Valve

Pump

METHOD

Air is pumped into the rubber bag till pressure in cuff is greater than pressure in artery even during heart's systole —
*   — Artery is then closed down during*
[At same time air is pushing up mercury
*   column in manometer.]*

SYSTOLE   and   DIASTOLE

By releasing valve on pump the pressure in cuff is gradually reduced till maximum pressure in artery just overcomes pressure in cuff — Some blood begins to spurt through during   SYSTOLE   — *Artery still closed during DIASTOLE*

At this point FAINT rhythmical TAPPING SOUNDS begin to be heard through STETHOSCOPE. The height of mercury in millimetres is taken as the SYSTOLIC Blood Pressure (e.g. 120 mm Hg)

Pressure in cuff is reduced still further till it is just less than the lowest pressure in artery towards the end of diastole (i.e. just before next heart beat) —
*   — Blood flow is unimpeded during*   SYSTOLE   and   DIASTOLE
The sounds stop. The height of mercury in the manometer at this point is taken as the DIASTOLIC Blood Pressure (e.g. about 80 mm Hg)

These values differ with SEX, AGE, EXERCISE, SLEEP, etc.

ELASTIC ARTERIES

The large <u>CONDUCTING ARTERIES</u> near the <u>HEART</u> are
ELASTIC ARTERIES

<u>SYSTOLE</u>
When <u>Ventricles</u> are <u>contracting</u>

<u>DIASTOLE</u>
When <u>Ventricles</u> are <u>relaxing</u>

pressure within
heart rises

pressure within
heart falls

AORTIC VALVES
are forced open

AORTIC VALVES
are pushed shut

Blood is forcibly
ejected into
AORTA

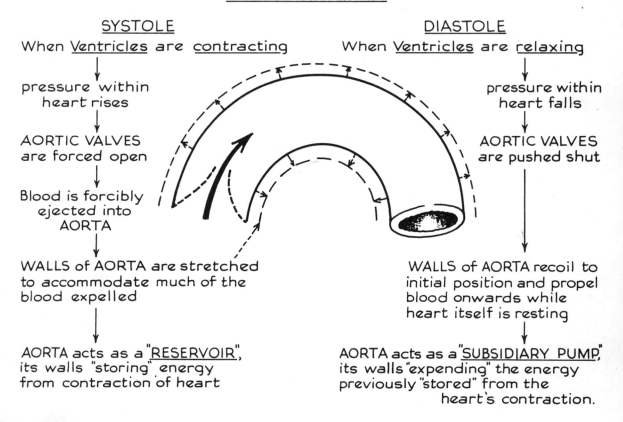

WALLS of AORTA are stretched
to accommodate much of the
blood expelled

WALLS of AORTA recoil to
initial position and propel
blood onwards while
heart itself is resting

AORTA acts as a "<u>RESERVOIR</u>",
its walls "storing" energy
from contraction of heart

AORTA acts as a "<u>SUBSIDIARY PUMP</u>",
its walls "expending" the energy
previously "stored" from the
heart's contraction.

As blood is pumped from the heart during systole, this distension
and increase in pressure which starts in the aorta passes along the
whole arterial system as a wave — the *pulse wave*.

The expansion and subsequent relaxation of the wall of the radial
artery can be felt as 'the pulse' at the wrist.

*A great increase in Blood Pressure can result if
these walls lose some of their elasticity with age or disease
and can no longer stretch readily to accommodate so much of
the heart's output during systole.*

94

NERVOUS REGULATION of MUSCULAR ARTERIES and ARTERIOLES

The Smooth Muscle in *Muscular Arteries* and *Arterioles* can contract or relax to alter calibre of the vessels, and thus adjust distribution of blood to meet the constantly changing needs of different parts of the body and also to maintain the normal blood pressure.

Highest Centres are probably in CEREBRAL CORTEX

....*HYPOTHALAMUS*

VASOMOTOR CENTRE in *MEDULLA OBLONGATA*

ARTERIOLE in SALIVARY GLANDS

PARASYMPATHETIC CRANIAL OUTFLOW

SYMPATHETIC OUTFLOWS

Stimulation of SYMPATHETIC (with accompanying inhibition of PARA-SYMPATHETIC) gives VASODILATATION

and

VASOCONSTRICTION *i.e. OVERALL EFFECT in body is one of VASOCONSTRICTION → increased Blood Pressure. [In these vessels vasodilatation seems to occur passively (blood pressure within vessels keeps them open) after reduction of sympathetic vasoconstrictor impulses.]*

ARTERIOLE in SKELETAL MUSCLE

ARTERIOLE in CORONARY SUPPLY to HEART   ?

ARTERIOLE in VISCERA   ?

ARTERIOLE in SKIN

ARTERIOLE in EXTERNAL GENITALIA

PARASYMPATHETIC SACRAL OUTFLOW

T1

Synapse with CONNECTOR NEURONES in Lateral Horns

EFFECTOR NEURONES leave the Sympathetic chain of Ganglia at all levels of Spinal Cord

L2

S2 3 4

Stimulation of PARASYMPATHETIC (with inhibition of SYMPATHETIC) appears to give VASODILATATION only in vessels to SALIVARY GLANDS and EXTERNAL GENITALIA.

Veins, and to some extent Capillaries, have similar nerve supply. Nervous Control of these vessels, however, seems to be less important than that of Muscular Arteries and Arterioles.

95

REFLEX and CHEMICAL REGULATION of ARTERIOLAR TONE – I

Afferent impulses are constantly reaching the VASOMOTOR CENTRE in the Medulla Oblongata in ingoing nerves from all parts of the body. These form the AFFERENT PATHWAYS for VASOMOTOR REFLEXES.

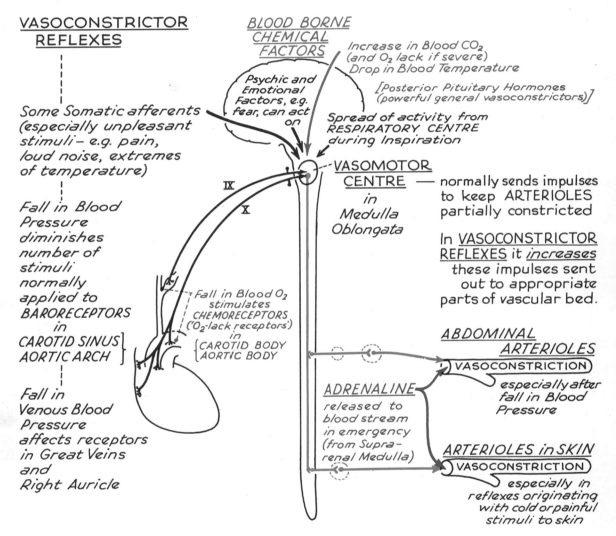

VASOCONSTRICTOR REFLEXES

Some Somatic afferents (especially unpleasant stimuli – e.g. pain, loud noise, extremes of temperature)

Fall in Blood Pressure diminishes number of stimuli normally applied to BARORECEPTORS in CAROTID SINUS AORTIC ARCH

Fall in Venous Blood Pressure affects receptors in Great Veins and Right Auricle

BLOOD BORNE CHEMICAL FACTORS

Psychic and Emotional Factors, e.g. fear, can act on

Increase in Blood CO_2 (and O_2 lack if severe) Drop in Blood Temperature

[Posterior Pituitary Hormones (powerful general vasoconstrictors)]

Spread of activity from RESPIRATORY CENTRE during Inspiration

IX

X

Fall in Blood O_2 stimulates CHEMORECEPTORS ('O_2-lack receptors') in [CAROTID BODY AORTIC BODY]

VASOMOTOR CENTRE in Medulla Oblongata — normally sends impulses to keep ARTERIOLES partially constricted

In VASOCONSTRICTOR REFLEXES it _increases_ these impulses sent out to appropriate parts of vascular bed.

ABDOMINAL ARTERIOLES

(VASOCONSTRICTION)

especially after fall in Blood Pressure

ADRENALINE released to blood stream in emergency (from Supra-renal Medulla)

ARTERIOLES in SKIN

(VASOCONSTRICTION)

especially in reflexes originating with cold or painful stimuli to skin

An _OVERALL VASOCONSTRICTION_, especially of abdominal vessels, gives rise in Blood Pressure.

REFLEX and CHEMICAL REGULATION of ARTERIOLAR TONE - II

VASODILATOR REFLEXES

Some Somatic afferents from skin (especially pleasant stimuli, -e.g. warmth)

REFLEXES for SALIVARY SECRETION

Rise in Blood Pressure stimulates BARORECEPTORS in CAROTID SINUS AORTIC ARCH

Rise in Blood O_2 reduces activity in CHEMORECEPTORS in { CAROTID BODY AORTIC BODY

BARORECEPTORS probably in Arteries of Mesentery, Veins in Abdomen and Thorax

VISCERAL AFFERENTS

ERECTILE TISSUE

Pudendal Nerve

BLOOD BORNE CHEMICAL FACTORS

Fall in Blood CO_2. Rise in Blood Temperature

Psychic and Emotional Factors can act on

VASOMOTOR CENTRE in Medulla Oblongata

ADRENALINE released to blood stream in emergency (from Supra-renal Medulla)

AXON REFLEX

Gentle stimulation of skin (i.e. by warmth or gentle massage)

In **VASODILATOR REFLEXES** vasoconstrictor 'tone' is reduced, i.e. fewer impulses are sent out to arterioles in appropriate regions of the body.

CORONARY ARTERIOLES (VASODILATATION)

Metabolites produced during tissue activity (e.g. CO_2, lactic acid) act directly on wall of small blood vessels

ARTERIOLES in SKELETAL MUSCLE (VASODILATATION)

ABDOMINAL ARTERIOLES (VASODILATATION) especially with Baroreceptor Reflex

ARTERIOLES in SKIN (VASODILATATION) in Reflexes originating with pleasant stimulation of skin

SKIN

An OVERALL VASODILATATION gives fall in Blood Pressure.

CAPILLARIES

Blood is distributed by *Arterioles* to *Capillaries*. Through the semi-permeable walls of *Capillaries* the sole purpose of the Circulating System is achieved.

FORCES determining EXCHANGE of WATER and ELECTROLYTES between
BLOOD ——— and ——— TISSUE FLUIDS

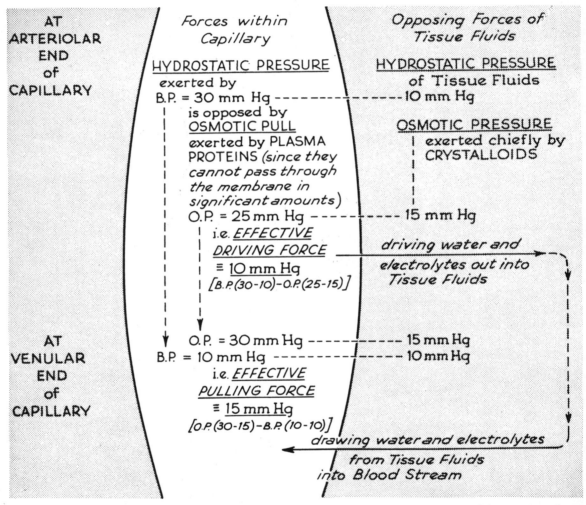

These exchanges in Capillaries result in a continuous "turn-over" and renewal of TISSUE FLUIDS.

VEINS: VENOUS RETURN

Capillaries unite to form *Veins* which *convey blood back to the heart.* By the time blood reaches the veins much of the force imparted to it by the heart's contraction has been spent.

<u>VENOUS RETURN</u> to the heart depends on various factors:-

GRAVITY -

from head and neck region.

"RESPIRATORY PUMP"

(a) *Intrathoracic pressure* is always lower than that in atmosphere. This *"negative pressure"* (which fluctuates with Respiratory Movements) exerts a *"suctioning"* pull which tends to draw column of blood upwards.

(b) Descent of *DIAPHRAGM* in Inspiration increases *Intra-abdominal pressure* and forces blood upwards in abdominal veins.

"MUSCULAR PUMP"

Contractions of skeletal muscles help to squeeze veins and move blood upwards.

"CARDIAC PUMP"

Residual force imparted by heart's contraction aids in forcing venous blood towards heart.

RIGHT AURICLE

DIAPHRAGM

"OVERFILLING" of the system (i.e. adequate BLOOD VOLUME) and ADEQUATE 'TONE' of smooth muscle in bloodvessel walls are essential.
(e.g. if too many vessels are dilated at any one time adequate venous return to the heart will not be maintained.)

ACTION of VALVES

Once the column of blood has moved up <u>UNIDIRECTIONAL VALVES</u> in larger veins prevent backflow.

CAPILLARIES

ARTERIAL PRESSURE

VENULES

BLOOD FLOW

The rate of blood flow varies in different parts of the vascular system. It is rapid in large vessels; slower in small vessels. *[The larger the* <u>*total*</u> *cross-sectional area represented the slower the rate of flow.]*

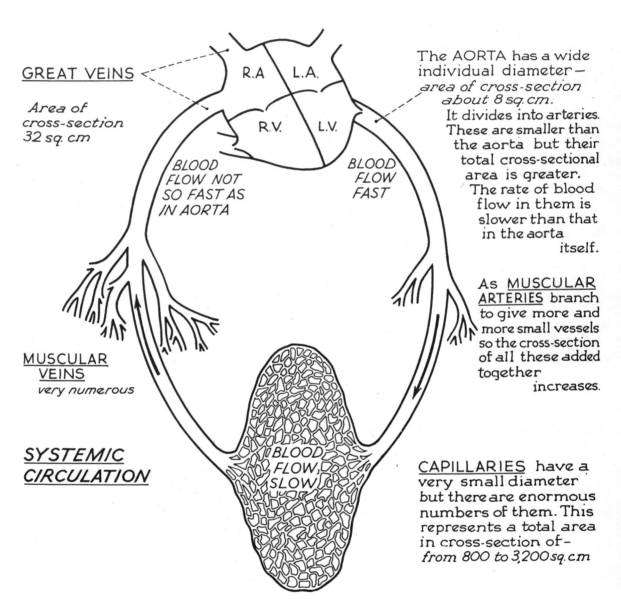

GREAT VEINS

Area of cross-section 32 sq. cm

R.A   L.A.

R.V.   L.V.

BLOOD FLOW NOT SO FAST AS IN AORTA

BLOOD FLOW FAST

The AORTA has a wide individual diameter— *area of cross-section about 8 sq. cm.*
It divides into arteries. These are smaller than the aorta but their total cross-sectional area is greater. The rate of blood flow in them is slower than that in the aorta itself.

As <u>MUSCULAR ARTERIES</u> branch to give more and more small vessels so the cross-section of all these added together increases.

MUSCULAR VEINS
very numerous

SYSTEMIC CIRCULATION

BLOOD FLOW, SLOW

CAPILLARIES have a very small diameter but there are enormous numbers of them. This represents a total area in cross-section of— *from 800 to 3,200 sq. cm*

100

PULMONARY CIRCULATION

BLOOD FLOW is faster in <u>PULMONARY CAPILLARIES</u> — Total cross-sectional area is smaller than that of *Systemic Capillaries.* Capillary pressure is less. This means that the osmotic 'pull' of the plasma proteins (25 mm Hg) exceeds the 'driving force' of the B.P. (5-10 mm Hg) along the whole length of the pulmonary capillary and fluid, therefore, cannot pass out into the alveoli.

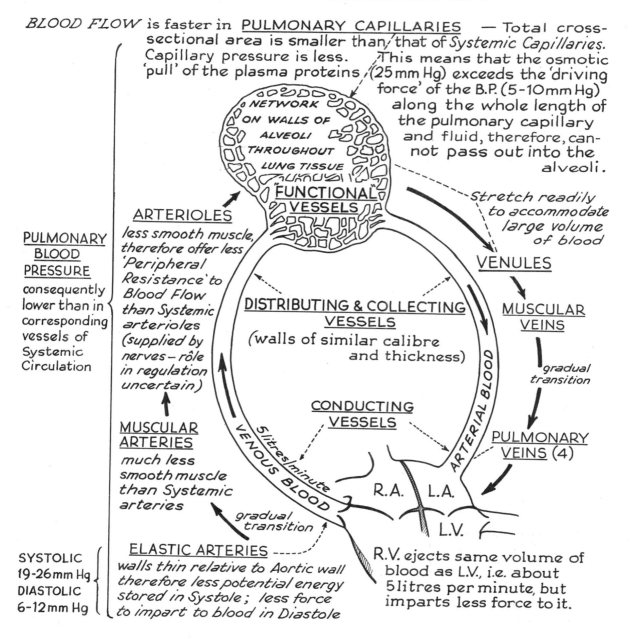

NETWORK ON WALLS OF ALVEOLI THROUGHOUT LUNG TISSUE

"FUNCTIONAL" VESSELS

<u>ARTERIOLES</u>
less smooth muscle, therefore offer less 'Peripheral Resistance' to Blood Flow than Systemic arterioles (supplied by nerves — rôle in regulation uncertain)

<u>PULMONARY BLOOD PRESSURE</u>
consequently lower than in corresponding vessels of Systemic Circulation

<u>MUSCULAR ARTERIES</u>
much less smooth muscle than Systemic arteries

Stretch readily to accommodate large volume of blood

<u>VENULES</u>

<u>MUSCULAR VEINS</u>

gradual transition

<u>DISTRIBUTING & COLLECTING VESSELS</u>
(walls of similar calibre and thickness)

<u>CONDUCTING VESSELS</u>

5 litres/minute VENOUS BLOOD

ARTERIAL BLOOD

<u>PULMONARY VEINS</u> (4)

R.A.   L.A.

L.V.

gradual transition

<u>ELASTIC ARTERIES</u>
walls thin relative to Aortic wall therefore less potential energy stored in Systole; less force to impart to blood in Diastole

SYSTOLIC 19-26 mm Hg
DIASTOLIC 6-12 mm Hg

R.V. ejects same volume of blood as L.V., i.e. about 5 litres per minute, but imparts less force to it.

A great dilatation of Pulmonary vessels occurs in exercise.

DISTRIBUTION of WATER and ELECTROLYTES in BODY FLUIDS

Water makes up about 70% of Adult Human Body —
— i.e about 46 litres in 70kg man.

Some lies outside the cells.
EXTRACELLULAR — 16 litres

A large part is within the cells.
INTRACELLULAR — 30 litres

BLOOD PLASMA	INTERSTITIAL FLUID
3·5 litres	*including* TISSUE FLUID 12 litres — bathes all Tissue Cells. Through this compartment all exchanges take place between

← BLOOD and TISSUES →

Cell membranes are freely permeable to WATER

H_2O ← Constant to-and-fro filtration between compartments → H_2O

← Passes freely from fluid of lower Osmotic Pressure to fluid of higher Osmotic Pressure →

This water contains substances dissolved in it.

PLASMA 55% of Total Blood Volume of 5·6 litres

Capillary Endothelium is freely permeable to ELECTROLYTES

Cell membranes are selectively permeable to ELECTROLYTES

← to-and-fro filtration →
in response to Osmotic Pressure Gradient

Na^+ ← → Na^+ (CHIEF CATION OF EXTRA-CELLULAR FLUID)

K^+
Mg^{++} } CHIEF CATIONS OF INTRACELLULAR FLUID.

(Protein⁻)
Cl^- → Cl^- { (CHIEF ANIONS OF EXTRA-CELLULAR FLUID)

$HPO_4^=$
$SO_4^=$
Protein⁻ } CHIEF ANIONS OF INTRACELLULAR FLUID.

HCO_3^- ← → HCO_3^- → HCO_3^- }

RED CELLS 45% of Total Blood Volume

H_2O
Cl^-
HCO_3^-
K^+
Hb^-

Interstitial Fluid Volume includes also small amounts of fluid in LYMPHATIC SYSTEM, CEREBROSPINAL FLUID, PERICARDIAL PLEURAL and PERITONEAL } SACS

[Areas in diagram are not proportional to actual volumes]

102

WATER BALANCE

In health the *total amount of body water (and salt)* is kept *reasonably constant* in spite of wide fluctuations in daily intake.

A **BALANCE** is struck between

FLUID INTAKE ——————and—————— FLUID OUTPUT

e.g. As LIQUID 1000 ml

In FOOD 1200 ml

TISSUE FLUIDS

PLASMA

SALIVA 1500 ml

ABSORPTION 2200 ml

GASTRIC JUICE 1500 ml

BILE 800 ml

PANCREATIC JUICE 1400 ml

INTESTINAL JUICE 1500 ml

REABSORPTION of SECRETIONS

150 ml

some NaCl, trace K⁺

IN FAECES

LUNGS 500 ml

BODY CELLS METABOLIC WATER produced per day 400 ml

400 ml

KIDNEYS Filtration 150,000 ml

NaCl Reabsorption 148,500 ml NaCl

SKIN (sweat glands) 450 ml

some NaCl

1500 ml

some NaCl K⁺ tubular secretion

IN SWEAT 'Seepage'   IN URINE

Except in Growth, Convalescence or Pregnancy, when new tissue is being formed, an INCREASE or DECREASE in INTAKE leads to an appropriate INCREASE or DECREASE in OUTPUT to maintain the BALANCE.

BLOOD

Blood is the specialized fluid tissue of the *TRANSPORT SYSTEM*. [*Specific Gravity, 1·055 – 1·065; pH, 7·3 - 7·4; Average amount, 5 litres, varies with body weight (about 7·7% of body weight).*

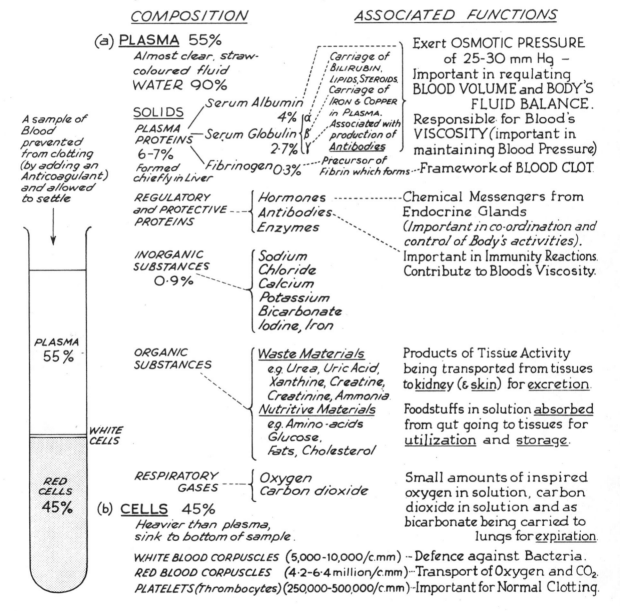

COMPOSITION

(a) PLASMA 55%
Almost clear, straw-coloured fluid
WATER 90%

SOLIDS
PLASMA PROTEINS 6-7%
Formed chiefly in Liver

Serum Albumin 4% α
Serum Globulin β
2·7% γ
Fibrinogen 0·3%

REGULATORY and PROTECTIVE PROTEINS
Hormones
Antibodies
Enzymes

INORGANIC SUBSTANCES 0·9%
Sodium
Chloride
Calcium
Potassium
Bicarbonate
Iodine, Iron

ORGANIC SUBSTANCES
Waste Materials
e.g. Urea, Uric Acid, Xanthine, Creatine, Creatinine, Ammonia.
Nutritive Materials
e.g. Amino-acids Glucose, Fats, Cholesterol

RESPIRATORY GASES
Oxygen
Carbon dioxide

(b) CELLS 45%
Heavier than plasma, sink to bottom of sample.

A sample of Blood prevented from clotting (by adding an Anticoagulant) and allowed to settle

PLASMA 55%

WHITE CELLS

RED CELLS 45%

ASSOCIATED FUNCTIONS

Carriage of BILIRUBIN, LIPIDS, STEROIDS.
Carriage of IRON & COPPER in PLASMA.
Associated with production of *Antibodies*
Precursor of Fibrin which forms

Exert OSMOTIC PRESSURE of 25-30 mm Hg –
Important in regulating BLOOD VOLUME and BODY'S FLUID BALANCE.
Responsible for Blood's VISCOSITY (important in maintaining Blood Pressure)
Framework of BLOOD CLOT.

Chemical Messengers from Endocrine Glands
(*Important in co-ordination and control of Body's activities*).
Important in Immunity Reactions.
Contribute to Blood's Viscosity.

Products of Tissue Activity being transported from tissues to kidney (& skin) for excretion.

Foodstuffs in solution absorbed from gut going to tissues for utilization and storage.

Small amounts of inspired oxygen in solution, carbon dioxide in solution and as bicarbonate being carried to lungs for expiration.

WHITE BLOOD CORPUSCLES (5,000-10,000/c.mm) - Defence against Bacteria.
RED BLOOD CORPUSCLES (4·2-6·4 million/c.mm) - Transport of Oxygen and CO_2.
PLATELETS (thrombocytes) (250,000-500,000/c.mm) - Important for Normal Clotting.

BLOOD COAGULATION

Blood does not normally clot within healthy blood vessels.
When *blood is shed* it undergoes a series of changes which end in CLOTTING.

The CLASSICAL THEORY of MORAWITZ recognized the interaction of 4 FACTORS

THROMBOPLASTIN Ⓘ —— acting on —— **PROTHROMBIN** Ⓘ

An enzyme thought to be liberated by damaged cells

A plasma protein — the precursor of Thrombin — formed in liver

Modern Views recognize *2 separate* sources of *THROMBOPLASTIN* in body — each results from a complex chain of reactions involving a number of *FACTORS.*

in presence of **CALCIUM** Ⓘ *ions which maintain electrolyte medium in which coagulation can take place*

In BLOOD PLASMA	In TISSUES
Hageman factor (XII)	*Factor V*
Plasma thromboplastin antecedent (XI)	
Factor V	*Tissue damage*
Factor VIII (Antihaemophilic globulin)	*Factor VII*
Platelets	
Factor IX (Christmas factor)	*Factor X (Stuart-Prower factor)*
Factor X (Stuart-Prower factor)	
Calcium	*Calcium*
INTRINSIC THROMBOPLASTIN	**EXTRINSIC THROMBOPLASTIN**

Ⓘ
THROMBIN which acts on **FIBRINOGEN**
(A SOLUBLE PROTEIN FORMED IN LIVER)

Insoluble FIBRIN deposited as fine threads to form FRAMEWORK of BLOOD CLOT

Platelets cling to intersections of fibrin threads

Both mechanisms appear to be necessary for PHYSIOLOGICAL HAEMOSTASIS
(i.e. cessation of bleeding when wound occurs) —— *also requires* ——▶ ADHESION and SHRINKAGE of CLOT

EXPRESSION of SERUM

FACTORS REQUIRED for NORMAL HAEMOPOIESIS

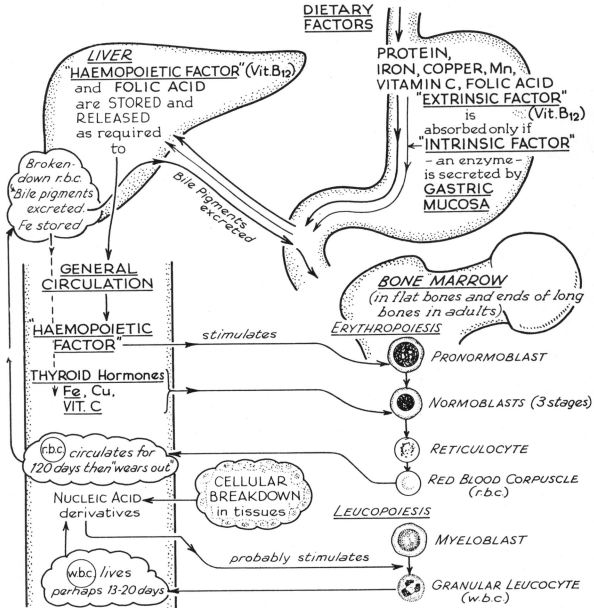

DIETARY FACTORS

LIVER
"HAEMOPOIETIC FACTOR" (Vit.B₁₂)
and FOLIC ACID
are STORED and
RELEASED
as required
to

PROTEIN,
IRON, COPPER, Mn,
VITAMIN C, FOLIC ACID
"EXTRINSIC FACTOR"
is (Vit.B₁₂)
absorbed only if
"INTRINSIC FACTOR"
- an enzyme -
is secreted by
GASTRIC MUCOSA

Broken-down r.b.c.
Bile pigments excreted.
Fe stored

Bile Pigments excreted

GENERAL CIRCULATION

BONE MARROW
(in flat bones and ends of long bones in adults)
ERYTHROPOIESIS

"HAEMOPOIETIC FACTOR" ──── stimulates ────

PRONORMOBLAST

THYROID Hormones
↓Fe, Cu, VIT. C

NORMOBLASTS (3 stages)

RETICULOCYTE

r.b.c. circulates for 120 days then "wears out"

CELLULAR BREAKDOWN in tissues

RED BLOOD CORPUSCLE (r.b.c.)

NUCLEIC ACID derivatives

LEUCOPOIESIS

MYELOBLAST

probably stimulates

w.b.c. lives perhaps 13-20 days

GRANULAR LEUCOCYTE (w.b.c.)

In health the number of r.b.c. and the amount of Hb in them remain fairly constant. Destruction of old red cells is balanced by formation of new.

HAEMOPOIESIS

In the adult the formed elements of the Blood Stream develop from primitive RETICULAR CELLS chiefly in RED BONE MARROW.

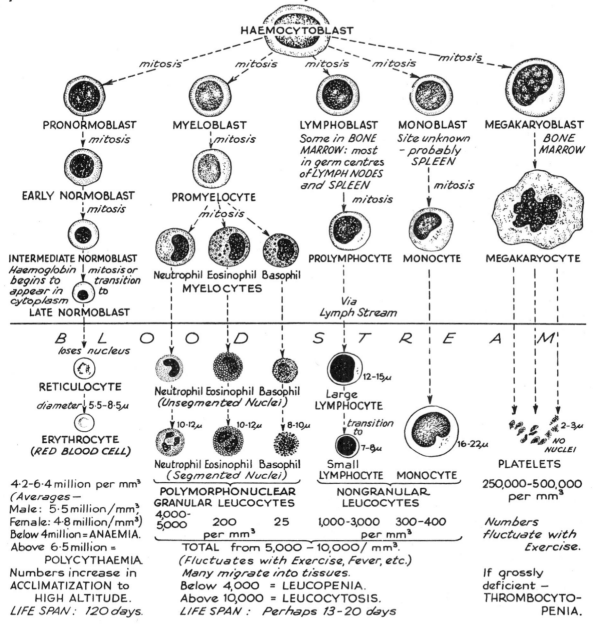

HAEMOCYTOBLAST

mitosis *mitosis* *mitosis* *mitosis* *mitosis*

PRONORMOBLAST **MYELOBLAST** **LYMPHOBLAST** **MONOBLAST** **MEGAKARYOBLAST**

mitosis *mitosis* Some in BONE MARROW: most in germ centres of LYMPH NODES and SPLEEN Site unknown – probably SPLEEN BONE MARROW

EARLY NORMOBLAST **PROMYELOCYTE** *mitosis* *mitosis*

mitosis *mitosis*

INTERMEDIATE NORMOBLAST Neutrophil Eosinophil Basophil **PROLYMPHOCYTE** **MONOCYTE** **MEGAKARYOCYTE**
Haemoglobin begins to appear in cytoplasm | *mitosis or transition to* MYELOCYTES

LATE NORMOBLAST

Via Lymph Stream

B L O O D S T R E A M

loses nucleus

RETICULOCYTE

diameter 5·5-8·5μ Neutrophil Eosinophil Basophil (Unsegmented Nuclei) Large LYMPHOCYTE 12-15μ

10-12μ 10-12μ 8-10μ *transition to* 7-8μ 16-22μ 2-3μ NO NUCLEI

ERYTHROCYTE (RED BLOOD CELL) Neutrophil Eosinophil Basophil (Segmented Nuclei) Small LYMPHOCYTE MONOCYTE **PLATELETS**

4·2-6·4 million per mm³ (Averages — Male: 5·5 million/mm³, Female: 4·8 million/mm³) Below 4 million = ANAEMIA. Above 6·5 million = POLYCYTHAEMIA. Numbers increase in ACCLIMATIZATION to HIGH ALTITUDE. *LIFE SPAN: 120 days.*

POLYMORPHONUCLEAR GRANULAR LEUCOCYTES 4,000-5,000

NONGRANULAR LEUCOCYTES

250,000-500,000 per mm³

| 200 | 25 | 1,000-3,000 | 300-400 |
| per mm³ | | per mm³ | |

TOTAL from 5,000 – 10,000/ mm³. (Fluctuates with Exercise, Fever, etc.) Many migrate into tissues. Below 4,000 = LEUCOPENIA. Above 10,000 = LEUCOCYTOSIS. *LIFE SPAN: Perhaps 13-20 days*

Numbers fluctuate with Exercise.

If grossly deficient – THROMBOCYTO-PENIA.

BLOOD GROUPS

There are present in the PLASMA of some individuals substances which can cause the AGGLUTINATION *(clumping together)* and subsequent HAEMOLYSIS *(breakdown)* of the RED BLOOD CELLS of some other individuals.

If such reactions follow BLOOD TRANSFUSION the two bloods are said to be INCOMPATIBLE.

Two FACTORS are involved in an AGGLUTINATION reaction :-
An AGGLUTINOGEN present in DONOR'S Red Blood Cell } e.g. A } or B}
A specific AGGLUTININ present in RECIPIENT'S Plasma } α } or β}

Obviously no such combination occurs naturally otherwise auto-agglutination would result.

In the ABO System

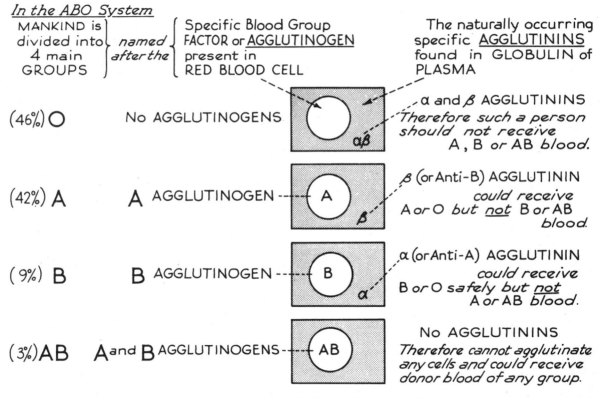

MANKIND is divided into 4 main GROUPS } *named after the* {	Specific Blood Group FACTOR or AGGLUTINOGEN present in RED BLOOD CELL		The naturally occurring specific AGGLUTININS found in GLOBULIN of PLASMA
(46%) O	No AGGLUTINOGENS		α and β AGGLUTININS *Therefore such a person should not receive* A , B or AB *blood.*
(42%) A	A AGGLUTINOGEN	A	β (or Anti-B) AGGLUTININ *could receive* A or O *but not* B or AB *blood.*
(9%) B	B AGGLUTINOGEN	B	α (or Anti-A) AGGLUTININ *could receive* B or O *safely but not* A or AB *blood.*
(3%) AB	A and B AGGLUTINOGENS	AB	No AGGLUTININS *Therefore cannot agglutinate any cells and could receive donor blood of any group.*

In Practice *it is important that the* DONOR's *cells should not be agglutinated by the* RECIPIENT's *plasma. Agglutination of Recipient's cells by Donor agglutinins is less likely to occur.*

[work.
There are sub-groups within these main groups. These have to be considered in blood transfusion

BLOOD GROUPS

To determine the Blood Group to which an individual belongs *TWO TEST SERA* only are required and the *RED BLOOD CELLS* to be grouped.

DROP of <u>GROUP A SERUM</u> & <u>GROUP B SERUM</u>
(on glass slide)   [N.B. These droplets are of PLASMA
 — <u>NOT</u> Red Blood Cells.
 i.e. only AGGLUTININS are present.]

To each test serum a saline suspension of RED BLOOD CELLS is added.   i.e. only AGGLUTINOGENS are added.

<u>GROUP A SERUM</u> | <u>GROUP B SERUM</u>

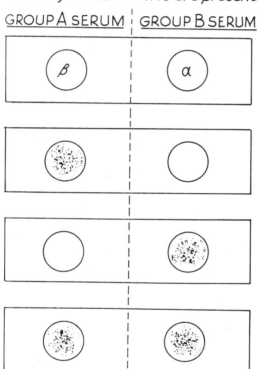

GROUP **O** Blood Cells give <u>NO</u> agglutination since <u>NO</u> AGGLUTINOGENS are present in these cells to be clumped by test sera AGGLUTININS.

GROUP **B** Blood Cells (B AGGLUTINOGEN present) give agglutination with GROUP **A** serum – since the specific Anti-B (β) AGGLUTININ is present in the first test serum.

GROUP **A** Blood Cells give agglutination with GROUP **B** serum — since the specific Anti-A (α) AGGLUTININ is present in this serum.

GROUP **AB** Blood Cells give agglutination with both test sera.

As well as determining Blood Group in this way, the Blood of Donor is always matched directly with Blood of Patient to avoid sub-group incompatibility.

AGGLUTINATION is usually visible under the microscope within a few minutes. The clumped cells look like grains of cayenne pepper in a clear liquid. If no agglutination occurs the fluid remains uniformly pink.

If the wrong blood is given to a patient, clumps of red blood cells may block small blood vessels in vital organs, e.g. lung or brain. The subsequent haemolysis (breakdown) of agglutinated cells may lead to Haemoglobin in the Urine and eventually to Kidney Failure and Death.

RHESUS FACTOR

Over 50 different BLOOD GROUP FACTORS have been demonstrated. Not all are important in blood transfusion work.

The _RHESUS FACTOR_ is important

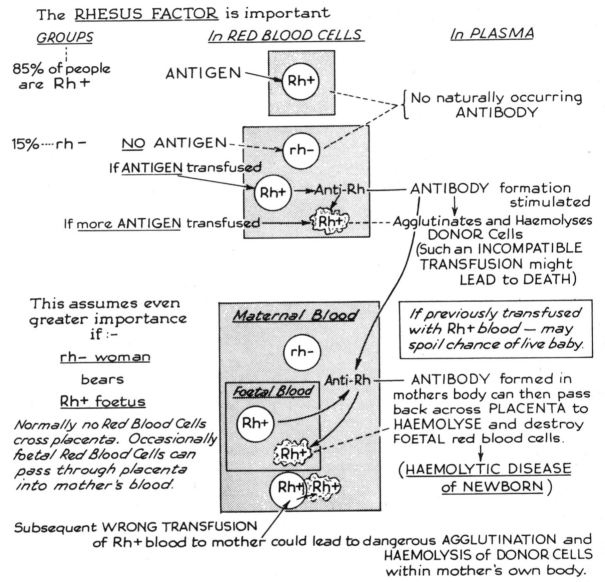

GROUPS   _In RED BLOOD CELLS_   _In PLASMA_

85% of people are Rh+

ANTIGEN — Rh+

No naturally occurring ANTIBODY

15%····rh−   NO ANTIGEN··· → rh−

If ANTIGEN transfused

Rh+ → Anti-Rh — ANTIBODY formation stimulated

If more ANTIGEN transfused ···· Rh+ ···· Agglutinates and Haemolyses DONOR Cells (Such an INCOMPATIBLE TRANSFUSION might LEAD to DEATH)

This assumes even greater importance if :−

rh− woman

bears

Rh+ foetus

Normally no Red Blood Cells cross placenta. Occasionally foetal Red Blood Cells can pass through placenta into mother's blood.

Maternal Blood

rh−

Foetal Blood

Rh+

Rh+

Anti-Rh

Rh+ Rh+

If previously transfused with Rh+ blood — may spoil chance of live baby.

— ANTIBODY formed in mothers body can then pass back across PLACENTA to HAEMOLYSE and destroy FOETAL red blood cells.

(HAEMOLYTIC DISEASE of NEWBORN)

Subsequent WRONG TRANSFUSION of Rh+ blood to mother could lead to dangerous AGGLUTINATION and HAEMOLYSIS of DONOR CELLS within mother's own body.

There are many sub-groups in this system.

110

INHERITANCE of RHESUS FACTOR

This BLOOD GROUP FACTOR is inherited on Mendelian principles.

HOMOZYGOUS Rh Positive FATHER	LACK of the FACTOR is a Mendelian recessive rh negative MOTHER

All Spermatozoa carry the FACTOR: *All Ova lack the FACTOR*

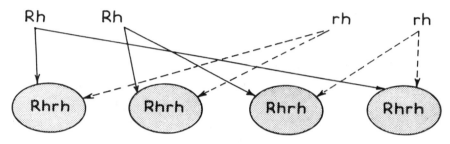

All children will be Rhrh, *i.e.* all carry the FACTOR and are Rh Positive. *i.e.* During pregnancy the rh negative woman will have a Rh Positive foetus.

HETEROZYGOUS Rh Positive FATHER	rh negative MOTHER

Half the Spermatozoa carry the FACTOR: *All Ova lack the FACTOR*

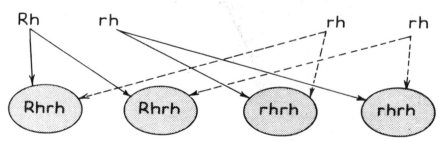

Possibility that some children will be Rhrh Positive, like the father; others rhrh negative like the mother.

LYMPHATIC SYSTEM

.All CELLS
are bathed by TISSUE FLUID.
This diffuses from CAPILLARIES.
Some returns to CAPILLARIES.
Some drains into blind-ending,
thin-walled LYMPHATICS.
It is then known as LYMPH
(similar to plasma but less protein).

A network of LYMPHATIC VESSELS drains tissue spaces throughout the body *(except in Central Nervous System)*. They unite to form LARGER and LARGER vessels ⟶ RIGHT LYMPHATIC DUCT and LEFT THORACIC DUCT ⟶ SUBCLAVIAN VEINS *(i.e. lymph is returned to the blood stream here)*

In the course of LARGER vessels, LYMPH is filtered through <u>LYMPH NODES</u>

AFFERENT LYMPHATICS– pour their LYMPH into RETICULAR FRAMEWORK of loose Sinus tissue.

MACROPHAGE cells ingest foreign material (e.g. carbon in lungs) or harmful bacteria.

LYMPH NODULES produce LYMPHOCYTES ⟶ in disintegration release PLASMA GLOBULIN (Ɣ Globulin -associated with antibody formation and immunity reactions).

Capsule of Fibrous Tissue

EFFERENT LYMPHATIC — receives lymph after its slow passage through node.

MOVEMENT of LYMPH towards HEART depends partly on compression of lymphatic vessels by muscles of limbs and partly on 'suction' created by movements of respiration. Valves within the vessels prevent backflow. The lymphatic tissue of the body forms an important part of the Body's Defence against invading agents such as PROTOZOA, BACTERIA, VIRUSES, or their poisonous TOXINS. These act as ANTIGENS stimulating ANTIBODY FORMATION —— which can subsequently destroy or neutralize the antigen.

SPLEEN

The Spleen is a vascular organ, weighing about 200 grams. It is situated in the left side of the abdomen behind the stomach and above the kidney.

FIBROUS TISSUE CAPSULE
with extensions into substance
of organ
TRABECULAE

SPLENIC ARTERY
and VEIN
and branches
TRABECULAR
ARTERY and VEIN

Near their
termination
ARTERIOLES
are surrounded
by collections of
LYMPHATIC TISSUE
("WHITE PULP") → produce
LYMPHOCYTES → added to
blood in its passage through spleen.

"RED PULP" — Framework of Reticular Tissue —
-acts as RESERVOIR for Blood.
(Some smooth muscle in Capsule and Trabeculae may
aid in expelling blood from reservoir to general
circulation)

PHAGOCYTIC CELLS of   Destroy worn out Red Blood
ELLIPSOID and other   Corpuscles, platelets and other
Reticulo-Endothelial Cells   (foreign) particles, etc.

It is difficult to trace further pathways of blood through
spleen and there is much disagreement about the exact
route taken.   Probably Arterioles link up with
VENOUS SINUSES (in the way shown).

-Blood appears to percolate through the narrow spaces
in the wall of the SINUS into the Red Pulp— and can get
back into Sinus in same way.

During foetal life the spleen forms red and white blood cells.
It is not apparently essential to life in the adult.

CEREBROSPINAL FLUID

COMPOSITION:- Cerebrospinal Fluid (C.S.F.) is similar to blood plasma but does not clot. It is a clear watery alkaline fluid. It contains salts, glucose, some urea and creatinine, very little protein and very few lymphocytes.
VOLUME:- 120-150 c.c. in man. SPECIFIC GRAVITY:- 1·005-1·008.

FORMATION

BRAIN is covered by 4 membranes

DURA MATER
  { Periosteal layer
  { Meningeal layer

ARACHNOID MATER

PIA MATER

The Subarachnoid Space - containing C.S.F. - links with Ventricles within the Brain.

ARACHNOID VILLI project into VENOUS SINUSES containing blood

C.S.F. from Foramen of Munro in Lateral Ventricle

III V.

×100

CHOROID PLEXUSES in Ventricles secrete C.S.F. continuously (500 c.c./day).

CIRCULATION
C.S.F. formed in Lateral Ventricles joins that from IIIrd. & IVth. Ventricles to circulate over surface of Brain and Spinal Cord in Subarachnoid space.

SPINAL CORD

DURA MATER

ARACHNOID MATER

PIA MATER

REABSORPTION through vascular tufts-ARACHNOID VILLI-into blood stream. [Effective forces:- HYDROSTATIC PRESSURE of C.S.F. - 120 mm H$_2$O - is greater than venous pressure in sinuses: it is aided by osmotic pull of Plasma Proteins within Plasma in returning C.S.F. to Blood Stream.]

FUNCTIONS of C.S.F.
1. Protective Covering for delicate brain tissues.
2. Alteration of Volume. can compensate for fluctuations in amount of blood within skull and thus keep total volume of cranial contents constant.
3. Exchange of Metabolic substances between Nerve Cells and C.S.F.
                    (i.e. it receives some waste products.)

114

CHAPTER 5

RESPIRATORY SYSTEM

RESPIRATORY SYSTEM

All living cells require to get OXYGEN from the fluid around them and to get rid of CARBON DIOXIDE to it.

INTERNAL RESPIRATION is the exchange of these gases between tissue cells and their fluid environment.

EXTERNAL RESPIRATION is the exchange of these gases *(oxygen and carbon dioxide)* between the body and the external environment.

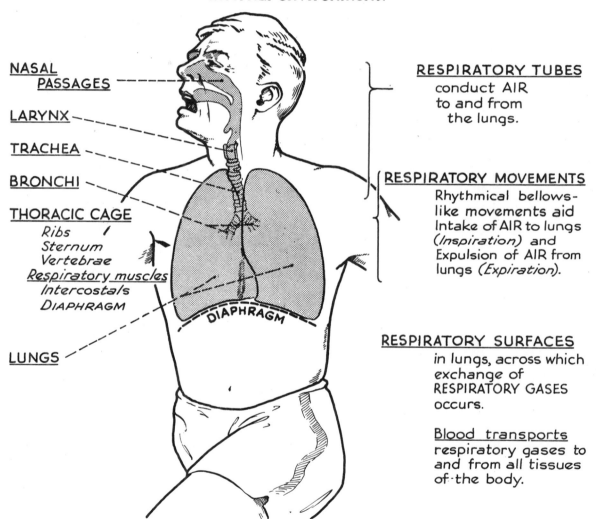

NASAL PASSAGES

LARYNX

TRACHEA

BRONCHI

THORACIC CAGE
Ribs
Sternum
Vertebrae
Respiratory muscles
Intercostals
DIAPHRAGM

DIAPHRAGM

LUNGS

RESPIRATORY TUBES
conduct AIR
to and from
the lungs.

RESPIRATORY MOVEMENTS
Rhythmical bellows-like movements aid Intake of AIR to lungs *(Inspiration)* and Expulsion of AIR from lungs *(Expiration)*.

RESPIRATORY SURFACES
in lungs, across which exchange of RESPIRATORY GASES occurs.

Blood transports respiratory gases to and from all tissues of·the body.

AIR CONDUCTING PASSAGES

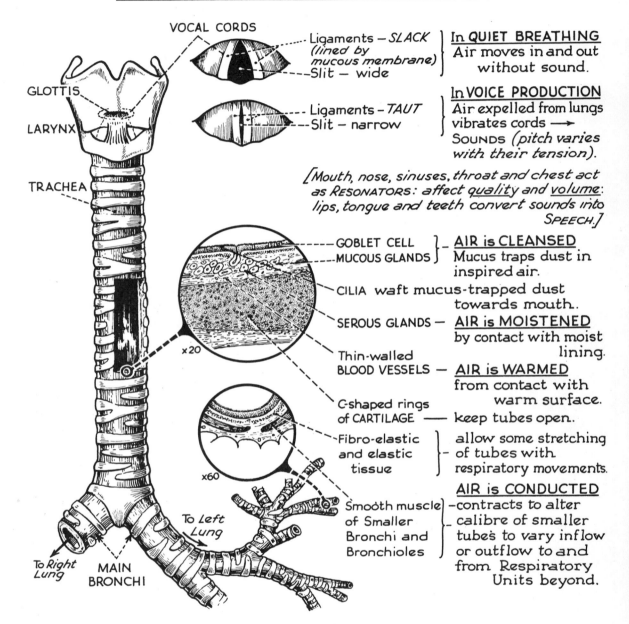

VOCAL CORDS

- Ligaments – *SLACK*
  *(lined by
  mucous membrane)*
- Slit – wide

GLOTTIS

LARYNX

- Ligaments – *TAUT*
- Slit – narrow

TRACHEA

In QUIET BREATHING
Air moves in and out
without sound.

In VOICE PRODUCTION
Air expelled from lungs
vibrates cords →
SOUNDS *(pitch varies
with their tension)*.

*[Mouth, nose, sinuses, throat and chest act
as RESONATORS: affect quality and volume:
lips, tongue and teeth convert sounds into
SPEECH.]*

- GOBLET CELL
- MUCOUS GLANDS

AIR is CLEANSED
Mucus traps dust in
inspired air.

x20

- CILIA waft mucus-trapped dust
  towards mouth.

- SEROUS GLANDS — **AIR is MOISTENED**
  by contact with moist
  lining.

- Thin-walled
  BLOOD VESSELS — **AIR is WARMED**
  from contact with
  warm surface.

- C-shaped rings
  of CARTILAGE —— keep tubes open.

- Fibro-elastic
  and elastic
  tissue

  allow some stretching
  of tubes with
  respiratory movements.

x60

- Smooth muscle
  of Smaller
  Bronchi and
  Bronchioles

AIR is CONDUCTED
–contracts to alter
calibre of smaller
tubes to vary inflow
or outflow to and
from Respiratory
Units beyond.

To Left
Lung

To Right
Lung

MAIN
BRONCHI

LUNGS: RESPIRATORY SURFACES

The Trachea and the Bronchial 'Tree' conduct Air down to the **RESPIRATORY SURFACES.**

There is no exchange of gases in these tubes.

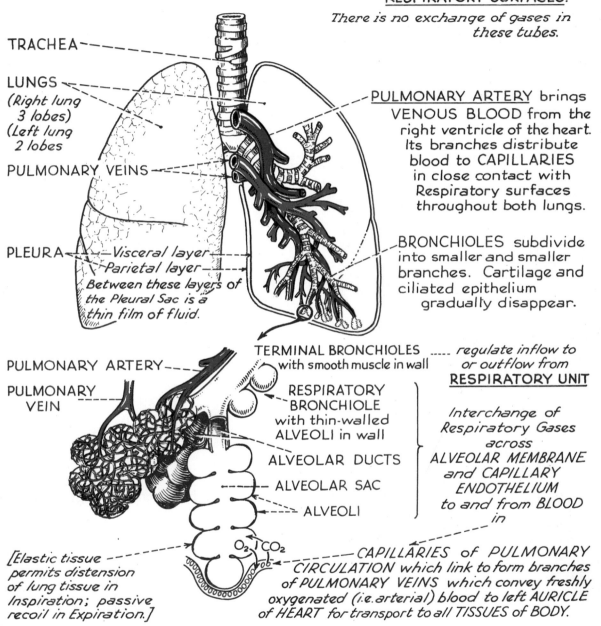

TRACHEA

LUNGS
(Right lung 3 lobes)
(Left lung 2 lobes

PULMONARY VEINS

PLEURA — *Visceral layer*
Parietal layer
Between these layers of the Pleural Sac is a thin film of fluid.

PULMONARY ARTERY

PULMONARY VEIN

PULMONARY ARTERY brings VENOUS BLOOD from the right ventricle of the heart. Its branches distribute blood to CAPILLARIES in close contact with Respiratory surfaces throughout both lungs.

BRONCHIOLES subdivide into smaller and smaller branches. Cartilage and ciliated epithelium gradually disappear.

TERMINAL BRONCHIOLES ..... with smooth muscle in wall

RESPIRATORY BRONCHIOLE with thin-walled ALVEOLI in wall

ALVEOLAR DUCTS

ALVEOLAR SAC

ALVEOLI

regulate inflow to or outflow from **RESPIRATORY UNIT**

Interchange of Respiratory Gases across ALVEOLAR MEMBRANE *and* CAPILLARY ENDOTHELIUM *to and from* BLOOD *in*

O_2 CO_2

[Elastic tissue permits distension of lung tissue in Inspiration; passive recoil in Expiration.]

CAPILLARIES *of* PULMONARY CIRCULATION *which link to form branches of* PULMONARY VEINS *which convey freshly oxygenated (i.e. arterial) blood to left* AURICLE *of* HEART *for transport to all TISSUES of BODY.*

119

THORAX

The Thorax (*or* chest) is the closed cavity which contains the LUNGS, HEART and Great Vessels.

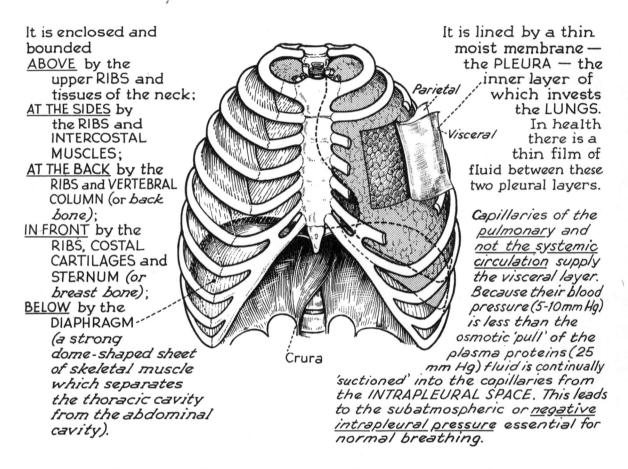

It is enclosed and bounded
<u>ABOVE</u> by the upper RIBS and tissues of the neck;
<u>AT THE SIDES</u> by the RIBS and INTERCOSTAL MUSCLES;
<u>AT THE BACK</u> by the RIBS and VERTEBRAL COLUMN (or *back bone*);
<u>IN FRONT</u> by the RIBS, COSTAL CARTILAGES and STERNUM (*or breast bone*);
<u>BELOW</u> by the DIAPHRAGM (*a strong dome-shaped sheet of skeletal muscle which separates the thoracic cavity from the abdominal cavity*).

Parietal

Visceral

Crura

It is lined by a thin moist membrane — the PLEURA — the inner layer of which invests the LUNGS. In health there is a thin film of fluid between these two pleural layers.

Capillaries of the <u>pulmonary and not the systemic circulation</u> supply the visceral layer. Because their blood pressure (5-10 mm Hg) is less than the osmotic 'pull' of the plasma proteins (25 mm Hg) fluid is continually 'suctioned' into the capillaries from the INTRAPLEURAL SPACE. This leads to the subatmospheric or <u>negative intrapleural pressure</u> essential for normal breathing.

DIMENSIONS of thoracic cage and the PRESSURE between pleural surfaces change rhythmically about 18-20 times a minute with the MOVEMENTS of RESPIRATION —— AIR MOVEMENT in and out of the lungs follows passively.

MECHANISM of BREATHING

The rhythmical changes in the CAPACITY of the Thorax are brought about by MUSCULAR ACTION. The changes in LUNG VOLUME with INTAKE or EXPULSION of air follow passively.

In NORMAL QUIET BREATHING

INSPIRATION

EXTERNAL INTERCOSTAL MUSCLES actively contract—
—ribs and sternum move upwards and outwards
—width of chest increases from side to side and from front to back.

DIAPHRAGM contracts—
—descends
—depth of chest increases.

CAPACITY of THORAX is INCREASED

↓

PRESSURE between PLEURAL SURFACES (already negative) is REDUCED from −2 to −6 mmHg (i.e. an increased "suction pull" is exerted on LUNG TISSUE)

ELASTIC TISSUE of LUNGS is STRETCHED

↓

LUNGS EXPAND to fill THORACIC CAVITY

↓

AIR PRESSURE within ALVEOLI is now less than atmospheric pressure

↓

AIR is sucked into ALVEOLI from ATMOSPHERE

EXPIRATION

EXTERNAL INTERCOSTAL MUSCLES relax—
—ribs and sternum move downwards and inwards
—width of chest diminishes.

DIAPHRAGM relaxes—
—ascends
—depth of chest diminishes.

CAPACITY of THORAX is DECREASED

↓

PRESSURE between PLEURAL SURFACES is INCREASED from −6 to −2 mm Hg (i.e. less pull is exerted on LUNG TISSUE)

ELASTIC TISSUE of LUNGS RECOILS

↓

AIR PRESSURE within ALVEOLI is now greater than atmospheric pressure

↓

AIR is forced out of ALVEOLI to ATMOSPHERE

In FORCED BREATHING

Muscles of Nostrils and round Glottis may contract to aid entrance of air to lungs.
Extensors of Vertebral Column may aid inspiration.
Muscles of Neck contract—
—move 1ST rib upwards
(and sternum upwards and forwards)

Internal Intercostals may contract—
—move ribs downwards more actively.
Abdominal muscles contract—
—actively aid ascent of diaphragm.

ARTIFICIAL RESPIRATION

Many methods aim at alternately DECREASING and INCREASING the CAPACITY of the THORAX 15-18 times per minute in imitation of the normal expiratory and inspiratory movements.

Two well-known methods illustrate the mechanism:-
[In each case mouth and throat must be clear of obstruction and tongue drawn forwards]

EXPIRATION   *INSPIRATION*

1. EVE'S ROCKING STRETCHER — uses the abdominal organs like a piston against the diaphragm.

EXPIRATION   *INSPIRATION*

2. SCHÄFER —— pressure applied to compress lower part of rib cage and so decrease capacity of thorax — *AIR MOVES OUT.*
On release of pressure thorax returns to previous position and capacity — *AIR SUCKED IN.*

In the MOUTH-to-MOUTH method the applicator pinches shut the patient's nostrils, seals his lips round the patient's mouth and blows in air [C]. When he removes his mouth the patient breathes out passively [D]. *[The residual oxygen in the applicator's own expired air is thus made available to patient]*

122

CAPACITY of LUNGS

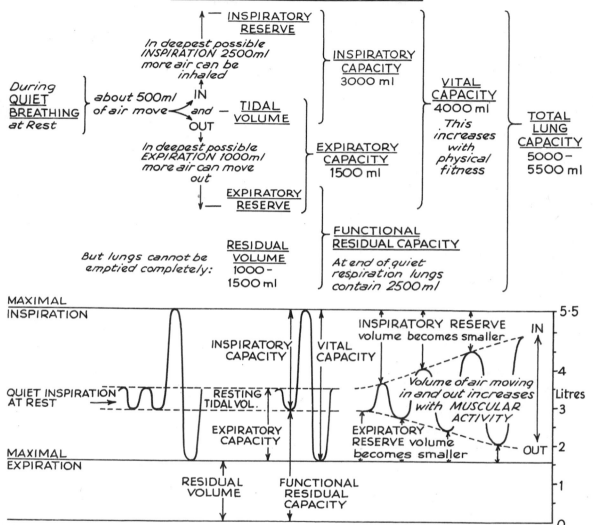

(After PAPPENHEIMER, J.R., et al (1950) Fed. Proc, 9, 602)

At rest a normal male adult breathes in and out about 16 times per minute. The amount of air breathed in per minute is therefore 500ml × 16, i.e. 8000ml or 8 litres – This is the RESPIRATORY MINUTE VOLUME or PULMONARY VENTILATION. In exercise it may go up to as much as 200 litres. These values are about 25% lower in women.

In deep breathing the volume of ATMOSPHERIC AIR INSPIRED with each inspiration and the amount which reaches the ALVEOLI increase.

COMPOSITION of RESPIRED AIR

In *QUIET BREATHING*

Of the
500 ml ATMOSPHERIC AIR
INSPIRED in a single inspiration
OXYGEN *makes up* 20·95%
NITROGEN ———— 79·01%
CARBON DIOXIDE— 0·04%

Of the AIR
EXPIRED in a single expiration
OXYGEN *makes up* 16·4%
NITROGEN ———— 79·6%
CARBON DIOXIDE— 4·0%

[N.B. Nitrogen is **not** given out by body.
The Volume of CO_2 given up by the
blood is slightly less than the
Volume of O_2 taken up by the blood.
The Volume of air expired is therefore
less than Volume inspired.
  Both contain the same *absolute*
amount of nitrogen.]

140 ml occupy the CONDUCTING
PASSAGES –"DEAD SPACE"
AIR.   This remains
unchanged in composition
since it is not in contact
with Respiratory Surfaces.

This represents a mixture of-
–"DEAD SPACE" AIR – air which
has moved out *unchanged*
from the CONDUCTING
              PASSAGES

and

360 ml (*together with 140 ml*
previously occupying
conducting passages)
reach the ALVEOLI to mix
with air already in
contact with
Respiratory Surfaces.
This mixture changes
composition as
it gives up
OXYGEN to the blood
and takes up
CARBON DIOXIDE
from the blood

ALVEOLAR AIR – air which
has been in contact with
Respiratory Surfaces and
has given up some OXYGEN
to the blood and taken up
CARBON DIOXIDE from it.
  OXYGEN ———— 13·8%
  NITROGEN ——— 80·7%
  CARBON DIOXIDE– 5·5%

With each EXPIRATION 140 ml
ALVEOLAR AIR move up to fill
Conducting passages.
This is drawn back into
ALVEOLI with start of next
INSPIRATION.

In *DEEP BREATHING* the composition of Air breathed out with each
          expiration comes closer to that of Alveolar Air.

MOVEMENT of RESPIRATORY GASES

A gas moves from an area where it is present at *higher concentration* to an area where it is present at *lower concentration*. The movement of gas molecules continues till the pressure exerted by them is the same throughout both areas. DRY Atmospheric Air *(at sea level)* has a pressure of 760 mm Hg

EXPIRED AIR

is laden with water vapour
exerting a pressure = **47 mm Hg**
O_2, 16·4% of (760−47) = **116·2 mm Hg**
CO_2, 4% of 713 = **28·5 mm Hg**

INSPIRED AIR

O_2, 20·95% of 760 = **159 mm Hg**
CO_2, 0·04% of 760 = **0·3 mm Hg**

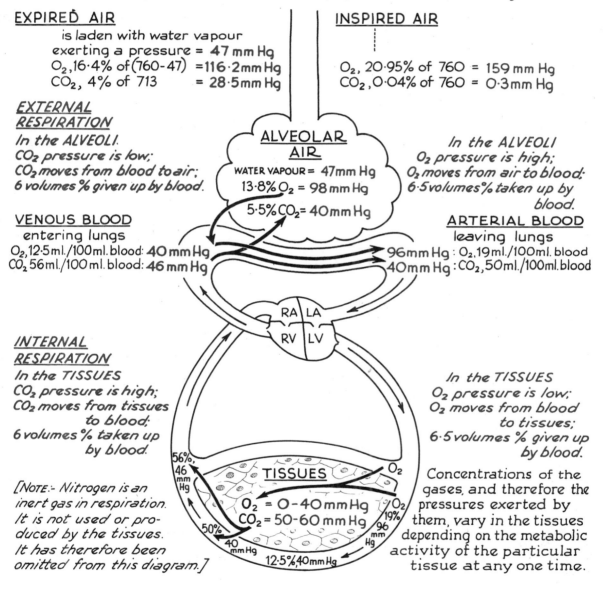

EXTERNAL RESPIRATION

In the ALVEOLI.
CO_2 pressure is low;
CO_2 moves from blood to air;
6 volumes % given up by blood.

In the ALVEOLI
O_2 pressure is high;
O_2 moves from air to blood:
6·5 volumes % taken up by blood.

ALVEOLAR AIR
WATER VAPOUR = 47 mm Hg
13·8% O_2 = 98 mm Hg
5·5% CO_2 = 40 mm Hg

VENOUS BLOOD

entering lungs
O_2, 12·5 ml./100 ml. blood: **40 mm Hg**
CO_2 56 ml./100 ml. blood: **46 mm Hg**

ARTERIAL BLOOD

leaving lungs
96 mm Hg : O_2, 19 ml./100 ml. blood
40 mm Hg : CO_2, 50 ml./100 ml. blood

RA LA
RV LV

INTERNAL RESPIRATION

In the TISSUES
CO_2 pressure is high;
CO_2 moves from tissues to blood:
6 volumes % taken up by blood.

In the TISSUES
O_2 pressure is low;
O_2 moves from blood to tissues;
6·5 volumes % given up by blood.

TISSUES
O_2 = 0−40 mm Hg
CO_2 = 50−60 mm Hg

56%, 46 mm Hg

50%
40 mm Hg

12·5%, 40 mm Hg

19% 96 mm Hg

[NOTE:- Nitrogen is an inert gas in respiration. It is not used or produced by the tissues. It has therefore been omitted from this diagram.]

Concentrations of the gases, and therefore the pressures exerted by them, vary in the tissues depending on the metabolic activity of the particular tissue at any one time.

DISSOCIATION of OXYGEN from HAEMOGLOBIN

The amount of O_2 taken up by HAEMOGLOBIN in the LUNGS or given up by OXYHAEMOGLOBIN in the TISSUES depends on the PARTIAL PRESSURE of the O_2 in the immediate environment.

It is also influenced by the PARTIAL PRESSURE of CO_2, by TEMPERATURE and by ACIDITY.

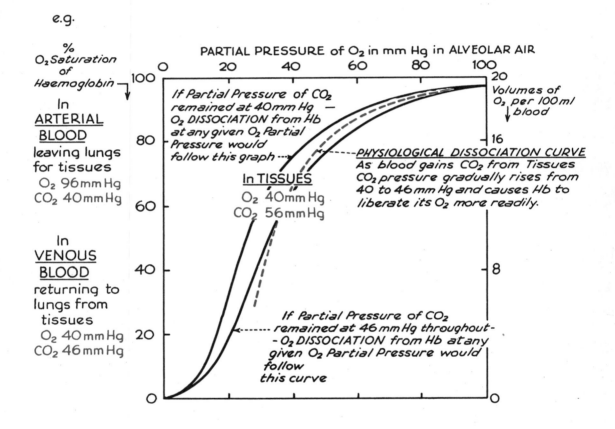

e.g.

%
O_2 Saturation
of
Haemoglobin →

In
ARTERIAL
BLOOD
leaving lungs
for tissues
O_2 96 mm Hg
CO_2 40 mm Hg

In
VENOUS
BLOOD
returning to
lungs from
tissues
O_2 40 mm Hg
CO_2 46 mm Hg

PARTIAL PRESSURE of O_2 in mm Hg in ALVEOLAR AIR

Volumes of O_2 per 100 ml blood

If Partial Pressure of CO_2 remained at 40mm Hg — O_2 DISSOCIATION from Hb at any given O_2 Partial Pressure would follow this graph

In TISSUES
O_2 40 mm Hg
CO_2 56 mm Hg

PHYSIOLOGICAL DISSOCIATION CURVE
As blood gains CO_2 from Tissues CO_2 pressure gradually rises from 40 to 46 mm Hg and causes Hb to liberate its O_2 more readily.

If Partial Pressure of CO_2 remained at 46 mm Hg throughout - O_2 DISSOCIATION from Hb at any given O_2 Partial Pressure would follow this curve

This effect of CO_2 partial pressure on dissociation of O_2 from Hb (the Böhr effect) is advantageous.

e.g. An increase in CO_2 partial pressure locally during tissue activity causes Hb to part more readily with its O_2 to the active tissues.

UPTAKE and RELEASE of CARBON DIOXIDE

CO_2 is carried by both red blood corpuscles and plasma. The bulk of it is carried as BICARBONATE formed chiefly in red cells and carried largely by plasma. The amount of CO_2 taken up by the blood in the tissues or released by the blood in the lungs depends on the PARTIAL PRESSURES of the CO_2 in the immediate environment and also on the state of the HAEMOGLOBIN or on the amounts of OXYGEN linked to Hb at any one time.

<u>In R.B.C.</u>   <u>In Plasma</u>

Hb ——————— and ———— Plasma Proteins act as *weak acids*.
KHb *(reduced Hb)* — and ———— Na Proteinate are salts of these acids.
K^+ ——————— and ———— Na^+ are easily displaced by stronger acids
                              (e.g. by H_2CO_3 formed from union of $CO_2 + H_2O$)
$KHbO_2$ *(oxygenated Hb)* is a *stronger* acid and
K^+ is <u>less</u> readily displaced from it by H_2CO_3.

PARTIAL PRESSURE of CO_2 in mm Hg

In
<u>VENOUS
BLOOD</u>
returning to
lungs from
tissues
O_2 40 mm Hg
CO_2 46 mm Hg

In
<u>ARTERIAL
BLOOD</u>
leaving lungs
for tissues
O_2 96 mm Hg
CO_2 40 mm Hg

If blood had only REDUCED Hb ——
when EXPOSED to increased CO_2 in TISSUES –
H_2CO_3 ($H_2O + CO_2$) + KHb → $KHCO_3$ + HHb *would proceed*
readily – i.e MORE CO_2 per mm³ could be carried.

ml CO_2
carried
by
100ml blood

PHYSIOLOGICAL UPTAKE
and RELEASE of CO_2
In blood, Hb is never completely
reduced or oxygenated.
Therefore as it loses O_2 to the
tissues HbO_2 decreases: more HHb
present → capacity to hold CO_2 gradually
increases as O_2 Partial Pressure decreases
and CO_2 Partial Pressure increases.

If blood had only OXYGENATED Hb ——
when EXPOSED to increased CO_2 in TISSUES –
H_2CO_3 + $KHbO_2$ → $KHCO_3$ + $HHbO_2$ would proceed
less readily –
i.e. LESS CO_2 per mm³ could be carried.

*i.e. The more OXYGEN the blood holds the less CO_2 it can hold and vice-versa.
This facilitates release of CO_2 in lungs and uptake of CO_2 in tissues.*

CARRIAGE and TRANSFER of OXYGEN and CARBON DIOXIDE

When **ARTERIAL BLOOD** is delivered by **SYSTEMIC CAPILLARIES** to the TISSUES it is exposed to:-

OXYGEN at very low tensions (1-40 mm Hg)

CARBON DIOXIDE at high tensions (50-60 mm Hg)

diffuses out of **PLASMA** where O_2 tension is higher

O_2 in SOLUTION

diffuses into **PLASMA** where CO_2 tension is lower. About 5% stays in SOLUTION — some combines with water.

RED BLOOD CELL

diffusion CO_2 ← CO_2

Chief PROTEIN of R.B.C. is Haemoglobin. It is attached chiefly to potassium.

$+$

H_2O ← H_2O

Carbonic anhydrase speeds up reaction in red blood cell

H_2CO_3

Proteins of Plasma are attached chiefly to sodium

HCO_3^- H^+ + Na Proteinate

i.e. salt of weak acid. Na^+ is displaced readily by H_2CO_3 (a stronger acid)

Oxygenated Hb

H_2CO_3

$KHbO_2$ + H^+ + HCO_3^-

Reduced Hb

HHb $\{Hb\}$ $\{H^+\}$ + $KHCO_3$ → HCO_3^- --- *diffuses from high to low concentration* → HCO_3^- → $NaHCO_3$ + H Proteinate

O_2

Na^+

→ K^+

Chloride shift [Cl⁻ ←---→ Cl⁻] exchanges to maintain ionic balance

$HHbCO_2$
(HHbNHCOOH)

CO_2

The bulk of CO_2 carried is combined with base as **BICARBONATE**

CO_2

A small amount of CO_2 carried by blood combines with a free amino group of HAEMOGLOBIN molecule to form **CARBAMINO COMPOUND**

During its passage through the Tissues blood gives up about 5-7 volumes % OXYGEN, i.e. its Hb is still up to 70% O_2 saturated.

The release of O_2 to tissues is speeded up by an increase in temperature or acidity such as occurs when tissues are active.

During its passage through the Tissues blood takes up about 4 to 6 volumes % CO_2.

Of the total CO_2 carried by blood, about 30% is carried in R.B.C., 70% in Plasma. Most of the BICARBONATE is *formed* in R.B.C. but the HCO_3^- diffusion gradient caused by carbonic anhydrase means that most of it is *carried* in plasma.

CARRIAGE and TRANSFER of OXYGEN and CARBON DIOXIDE

When <u>VENOUS BLOOD</u> flows through the <u>PULMONARY CAPILLARIES</u> it is exposed to:-

<u>OXYGEN</u> at high pressures
in ALVEOLAR AIR (98 mm Hg)

<u>CARBON DIOXIDE</u> at low pressures
(40 mm Hg)

diffuses into

<u>PLASMA</u> where O_2 tension is lower

diffuses out of

<u>PLASMA</u> where CO_2 tension is higher. In SOLUTION – released mainly from BICARBONATE in R.B.C.; some from Plasma

O_2 in SOLUTION

RED BLOOD CELL

CO_2
+
H_2O

CO_2
+
H_2O

Carbonic anhydrase accelerates reaction in R.B.C.

H_2CO_3

$HCO_3^- + H^+ + Na\ Proteinate$

Oxygenated Haemoglobin

H_2CO_3

$KHbO_2 + H^+ + HCO_3^-$

Reduced Hb

$HHb \left\{ \begin{array}{l} Hb^- \\ H^+ \end{array} \right.$ O_2 $+ KHCO_3$

HCO_3^- K^+

HCO_3^-

$NaHCO_3 + H\ Proteinate$

Na^+

$[Cl^- \text{------} Cl^-]$
Chloride shift

$HHbCO_2$ CO_2

CO_2 *From*
<u>CARBAMINO COMPOUND</u>

As blood passes through capillaries of lungs it takes up approximately 5 to 7 volumes % of OXYGEN. O_2 combines with Haemoglobin (Hb) molecule. It becomes about 95–97% saturated with Oxygen.

As blood passes through capillaries of lungs it gives up approximately 4 to 6 volumes % of CARBON DIOXIDE. A small amount is released from combination with the free amino group in Haemoglobin molecule – Carbamino compound. Most comes from BICARBONATE in R.B.C. and Plasma by processes indicated in diagram.

NERVOUS CONTROL of RESPIRATORY MOVEMENTS

Normal Respiratory Movements are Involuntary. They are carried out automatically (i.e. without conscious control) through the rhythmical discharge of nerve impulses from CONTROLLING CENTRES in the BRAIN.

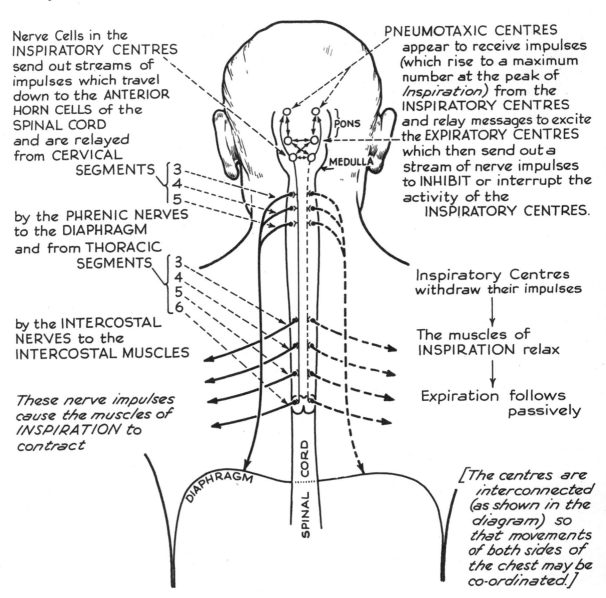

Nerve Cells in the INSPIRATORY CENTRES send out streams of impulses which travel down to the ANTERIOR HORN CELLS of the SPINAL CORD and are relayed from CERVICAL SEGMENTS 3 4 5

by the PHRENIC NERVES to the DIAPHRAGM and from THORACIC SEGMENTS 3 4 5 6

by the INTERCOSTAL NERVES to the INTERCOSTAL MUSCLES

These nerve impulses cause the muscles of INSPIRATION to contract

PNEUMOTAXIC CENTRES appear to receive impulses (which rise to a maximum number at the peak of *Inspiration*) from the INSPIRATORY CENTRES and relay messages to excite the EXPIRATORY CENTRES which then send out a stream of nerve impulses to INHIBIT or interrupt the activity of the INSPIRATORY CENTRES.

PONS

MEDULLA

Inspiratory Centres withdraw their impulses

↓

The muscles of INSPIRATION relax

↓

Expiration follows passively

[The centres are interconnected (as shown in the diagram) so that movements of both sides of the chest may be co-ordinated.]

DIAPHRAGM

SPINAL CORD

CHEMICAL REGULATION of RESPIRATION

The activity of the Respiratory Centres is affected by the chemistry and temperature of the blood supplying them.

The most important factor in regulating the activity of the respiratory centres is the level of CARBON DIOXIDE in the blood passing through them. —— This leads to an alteration in outgoing impulses to respiratory muscles.

Increase in CO_2
Increase in H^+ concentration } stimulates RESPIRATORY CENTRES --→ *increases depth of breathing (and eventually increases rate of breathing).*

Fall in blood CO_2 depresses ------ -→ *shallow breathing.*

Increase in temperature of blood -- -→ *quickens but does not deepen respiration.*

Cooling of blood ------- -→ *slows respiration.*

Normally respiration is little affected by slight changes in O_2 content of blood:

Severe lack of O_2 depresses respiratory centre

CHEMO-REFLEXES:
In addition to the direct effect of CO_2 and O_2 on centre

Lack of O_2 ---- *stimulates* CHEMORECEPTORS *(oxygen-lack receptors) in* CAROTID BODY *and* AORTIC BODY } -- *reflexly stimulates respiration*
e.g. as at low atmospheric pressure (high altitude)

NOTE:- *These reflexes are usually powerful enough to override the direct depressant action of lack of O_2 on respiratory centres themselves.*

The CHEMICAL and NERVOUS means of regulating the activity of RESPIRATORY CENTRES act together to adjust rate and depth of breathing to the needs of the body.

VOLUNTARY and REFLEX FACTORS in the REGULATION of RESPIRATION

Although fundamentally automatic and regulated by chemical factors in the blood, ingoing impulses from many parts of the body also modify the activity of the RESPIRATORY CENTRES and consequently alter the outgoing impulses to the Respiratory muscles to co-ordinate RHYTHM, RATE or DEPTH of breathing with other activities of the body.

Impulses from **HIGHER CENTRES** – PSYCHIC and EMOTIONAL INFLUENCES
{
Voluntary alterations in Breathing.
Interruptions of expiration in SPEECH and SINGING.
Deep inspiration then short spasmodic expirations in LAUGHTER and WEEPING.
Prolonged expiration in SIGHING.
Deep inspiration with mouth open in YAWNING.
Slow shallow breathing in SUSPENSE and CONCENTRATION.
Rapid breathing in FEAR and EXCITEMENT.
}

REFLEX Influences

e.g. PUNGENT ODOURS *irritating nerve endings in Nasal mucosa.*

BOLUS of FOOD *contacting Pharynx.*

IRRITANT *contacting Larynx, Trachea*

PAINFUL, HOT, COLD *stimuli to nerve endings in skin*

Stretch-proprioceptors *in INTERCOSTAL muscles, DIAPHRAGM, ABDOMINAL muscles*

Decrease (↓) in *blood pressure*

Increase (↑) in *blood pressure*

BARORECEPTORS in CAROTID sinus and AORTIC sinus

↓O₂

CHEMORECEPTORS in CAROTID BODY and AORTIC BODY

REFLEX alterations in Respiratory movements

→ *short inspiration; forced expirations with GLOTTIS open in SNEEZING.*

→ *inhibition of respiration during SWALLOWING.*

→ *short inspiration; series of forced expirations with GLOTTIS closed (high pressure created in air passages); GLOTTIS opens suddenly; blast of air carries out irritant material in COUGHING.*

→ *sharp inspiration after sudden PAIN or COLD; increasing rate and depth of breathing with HEAT.*

→ *spasmodic contractions of diaphragm with GLOTTIS closed in HICCOUGHING.*

→ *respiration stimulated.*

→ *respiration depressed.*

In Normal Breathing respiratory rate and rhythm are influenced rhythmically by the **HERING-BREUER REFLEX.**

| Distension of ALVEOLI at end of INSPIRATION | stimulates STRETCH RECEPTORS in bronchioles | Stream of ingoing impulses passes along VAGUS nerves to depress INSPIRATORY CENTRES | Withdrawal of outgoing impulses to Respiratory muscles →EXPIRATION. |

CHAPTER 6

EXCRETORY SYSTEM

133

EXCRETORY SYSTEM

Together with the *Respiratory System* and the *Skin*, the KIDNEYS are the chief *Excretory organs* of the body.

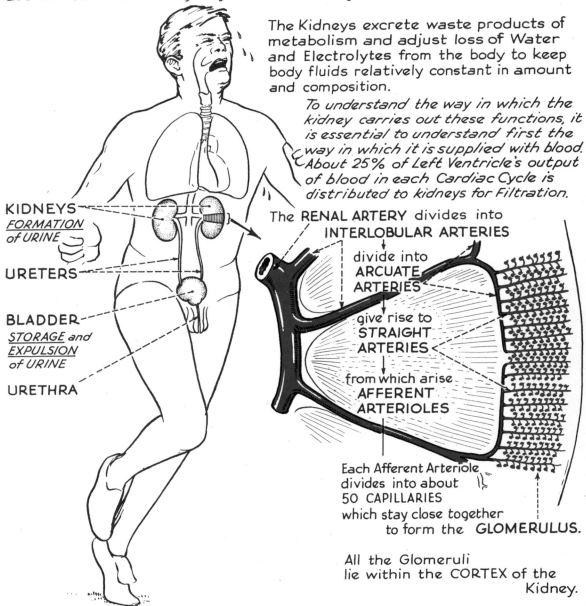

The Kidneys excrete waste products of metabolism and adjust loss of Water and Electrolytes from the body to keep body fluids relatively constant in amount and composition.

To understand the way in which the kidney carries out these functions, it is essential to understand first the way in which it is supplied with blood. About 25% of Left Ventricle's output of blood in each Cardiac Cycle is distributed to kidneys for Filtration.

KIDNEYS
FORMATION of URINE

URETERS

BLADDER
STORAGE and EXPULSION of URINE

URETHRA

The **RENAL ARTERY** divides into **INTERLOBULAR ARTERIES**

divide into **ARCUATE ARTERIES**

give rise to **STRAIGHT ARTERIES**

from which arise **AFFERENT ARTERIOLES**

Each Afferent Arteriole divides into about 50 CAPILLARIES which stay close together to form the GLOMERULUS.

All the Glomeruli lie within the CORTEX of the Kidney.

KIDNEY

Each Kidney contains approximately one million microscopic units — *NEPHRONS* – which form URINE.

In the Renal Corpuscle urine formation starts with FILTRATION of the Blood

Each AFFERENT ARTERIOLE leads to a capillary tuft— the GLOMERULUS Surrounding this tuft is the closed end— BOWMAN'S CAPSULE—

–of a long tortuous RENAL TUBULE which has various parts—

The blood leaving the capillary tuft flows in an EFFERENT ARTERIOLE which divides to form a 2nd CAPILLARY network round the tubules of its own nephron. These capillaries finally converge into a VEIN.

FIBROUS TISSUE CAPSULE

1ST CONVOLUTED TUBULE

2ND CONVOL. TUBULE

In the tubules urine formation is completed by REABSORPTION to the Blood Stream of essential materials from the Filtrate and by SECRETION and SYNTHESIS of other waste materials

CORTEX

MEDULLA

RENAL ARTERY and VEIN

PELVIS

PYRAMID

DESCENDING AND ASCENDING LIMBS OF LOOP of HENLE

COLLECTING TUBULES empty formed urine into the PELVIS of the KIDNEY.

CALYX

URETER

FORMATION of URINE — 1. FILTRATION

About 25% of the left ventricle's total output of blood in each cardiac cycle is distributed through the Renal Arteries to the Kidneys for FILTRATION.

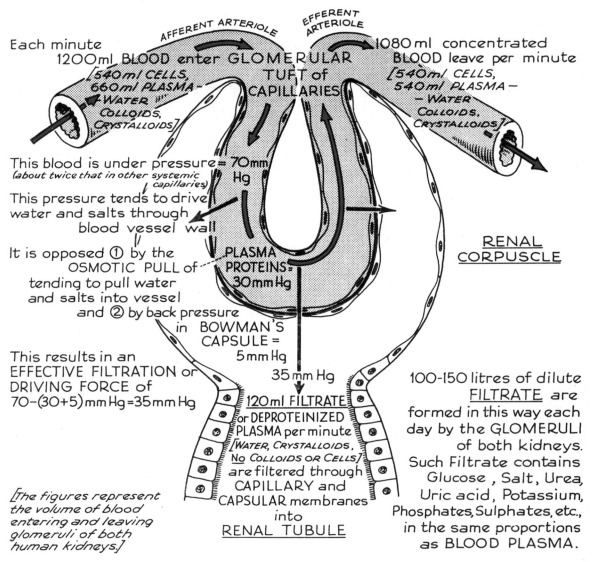

AFFERENT ARTERIOLE

EFFERENT ARTERIOLE

Each minute 1200ml BLOOD enter GLOMERULAR TUFT of CAPILLARIES [540ml CELLS, 660ml PLASMA — WATER, COLLOIDS, CRYSTALLOIDS]

1080ml concentrated BLOOD leave per minute [540ml CELLS, 540ml PLASMA — WATER, COLLOIDS, CRYSTALLOIDS]

This blood is under pressure = 70mm Hg (about twice that in other systemic capillaries)

This pressure tends to drive water and salts through blood vessel wall

It is opposed ① by the OSMOTIC PULL of PLASMA PROTEINS: 30mm Hg tending to pull water and salts into vessel and ② by back pressure in BOWMAN'S CAPSULE = 5mm Hg

RENAL CORPUSCLE

This results in an EFFECTIVE FILTRATION or DRIVING FORCE of 70–(30+5) mm Hg = 35 mm Hg

35 mm Hg

120ml FILTRATE or DEPROTEINIZED PLASMA per minute [WATER, CRYSTALLOIDS, No COLLOIDS OR CELLS] are filtered through CAPILLARY and CAPSULAR membranes into RENAL TUBULE

[The figures represent the volume of blood entering and leaving glomeruli of both human kidneys.]

100-150 litres of dilute FILTRATE are formed in this way each day by the GLOMERULI of both kidneys. Such Filtrate contains Glucose, Salt, Urea, Uric acid, Potassium, Phosphates, Sulphates, etc., in the same proportions as BLOOD PLASMA.

The Glomerular membrane acts as a simple FILTER —
— i.e. no energy is used up by the cells in filtration.

FORMATION of URINE–2. CONCENTRATION

As it passes along the TUBULE the FILTRATE is CONCENTRATED and essential substances are CONSERVED. The Tubular Epithelium reabsorbs water and selected essential materials into the blood stream.

GLOMERULAR FILTRATE 120 ml/min

BLOOD LEAVING GLOMERULI 1080 ml/minute

BLOOD LEAVING KIDNEY 1199 ml/minute

Most of the WATER is reabsorbed

Total WATER reabsorbed = 119 ml
of this—
80% is reabsorbed in proximal tubule
20% is reabsorbed in distal tubule and partly in collecting ducts

DISCARDED materials pass out in the URINE –1ml/minute passes into EXCRETORY or COLLECTING DUCTS

ESSENTIAL or "HIGH THRESHOLD" Substances
  e.g. GLUCOSE
  SODIUM CHLORIDE
  and other essential salts
  AMINO ACIDS
  VITAMIN C
–are all reabsorbed to PERITUBULAR CAPILLARIES

–unless their concentration in the blood is very high

MEDIUM THRESHOLD
  e.g. POTASSIUM –
some reabsorbed.
(can also be secreted by Tubular epithelium of distal tubule).

LOW THRESHOLD
  e.g. UREA, PHOSPHATES, URIC ACID –
only small amounts reabsorbed.

NO THRESHOLD
e.g. CREATININE, SULPHATES–
not reabsorbed at all.
(Tubular epithelium can also secrete some CREATININE into the tubules)

REACTION is CHANGED
e.g.
ALKALINE PHOSPHATE changed to ACID PHOSPHATE to conserve body's base and to keep blood and body fluids alkaline. Tubular epithelium forms AMMONIA which combines with acids → excreted as Ammonium Salts.

[Ascending limb of Loop of Henle is impermeable to water (see next page)]

PELVIS of KIDNEY

1–1½ litres of URINE are excreted per day.

In concentrating the filtrate and in secretion of certain substances, the tubular epithelial cells use energy, i.e. they do work — much of it against an osmotic gradient.

FORMATION of URINE:
MECHANISM of WATER REABSORPTION

Water is <u>not</u> actively reabsorbed by tubular cells. Its movements are determined <u>passively</u> by the <u>osmotic gradient</u> set up by solutes — chiefly by the <u>sodium</u> salts.

In CORTEX:-
PROXIMAL TUBULE
Filtrate isotonic

Active transport of Na^+, H_2O follows passively, *(drawn by 'osmotic pull')*

DISTAL TUBULE
Filtrate hypotonic

Na^+ active reabsorption under influence of **ALDOSTERONE**. H_2O passively

Contents here are **ISOTONIC** (i.e. same O.P. as plasma)

CAPILLARY *arteriolar end*

DESCENDING LIMB *Permeable to H_2O*

LOOP of HENLE

ASCENDING LIMB *Impermeable to H_2O*

CAPILLARY *venular end*

COLLECTING DUCT
made permeable to H_2O by **ADH**. URINE flowing towards tip of papilla becomes *progressively more concentrated* due to osmotic withdrawal of H_2O.

In MEDULLA:-
CELLS, TISSUE FLUIDS and BLOOD become *progressively more concentrated* as they pass towards tip of papilla, i.e. O.P. rises progressively. This *osmotic gradient* is set up and maintained by the 'build up' of *sodium salts* in the region. This in turn is made possible by the anatomical arrangement of the <u>LOOP of HENLE</u> and its accompanying <u>CAPILLARIES</u>.

TISSUE FLUIDS

Na^+ diffusion

Na^+ diffusion (Cl^-)

$\leftarrow Na^+ \leftarrow$ $(Cl^-$ follows$)$

Na^+ active transport

[water cannot follow since wall here impermeable to it] i.e. contents become *more dilute* (*less* Na^+Cl^- present)

Na^+ trapped in region of the loops

$\leftarrow Na^+$

becoming more concentrated (due to additional Na^+(Cl^-))

becoming more concentration (∵ added Na^+(Cl^-))

HYPERTONIC

becoming more dilute (∵ H_2O added)

O.P. — H_2O →

O.P. — H_2O →

O.P. — H_2O →

HYPERTONIC (i.e. more concentrated than plasma but same O.P. as tissues at tip of papilla.

[Without ADH, cells lining collecting duct stay impermeable to water and a hypotonic urine can be excreted.]

HYPERTONIC

URINE

PELVIS of KIDNEY

These form a "<u>COUNTER-CURRENT SYSTEM</u>" (i.e <u>outflowing</u> fluid flows counter to (yet near) <u>inflowing</u>). Active transport of sodium ions out of the ascending limb (without accompanying loss of water through its 'water-tight' walls) and its subsequent diffusion into the descending limb is the key to the chain of events shown.

This mechanism permits maximal {reabsorption of water} {concentration of urine} with minimal expenditure of energy by tubular cells.

The "CLEARANCE" of INULIN in the NEPHRON

The rate of Glomerular Filtration can be measured by injecting a substance which is known to be *filtered off* by the Renal Corpuscle but *not secreted* or *reabsorbed* by the Tubular epithelium.
INULIN is such a substance. It is *filtered* from the Glomerular Capillaries.

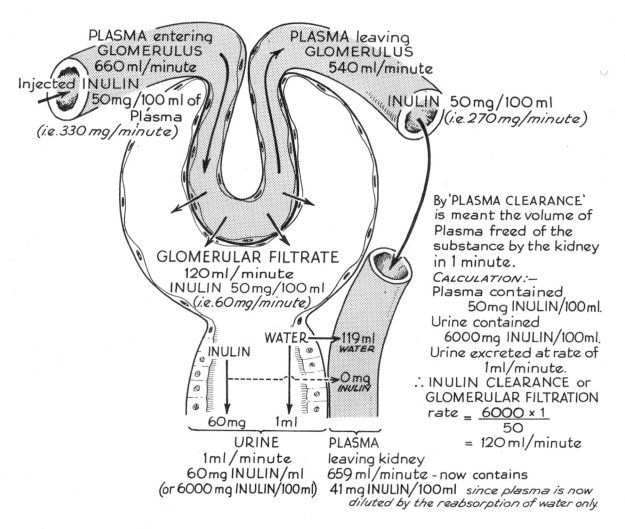

PLASMA entering
GLOMERULUS
660 ml/minute
Injected INULIN
50 mg/100 ml of
Plasma
(*i.e. 330 mg/minute*)

PLASMA leaving
GLOMERULUS
540 ml/minute
INULIN 50 mg/100 ml
(*i.e. 270 mg/minute*)

GLOMERULAR FILTRATE
120 ml/minute
INULIN 50 mg/100 ml
(*i.e. 60 mg/minute*)

INULIN

WATER → 119 ml
WATER

→ 0 mg
INULIN

60 mg   1 ml

URINE
1 ml/minute
60 mg INULIN/ml
(or 6000 mg INULIN/100 ml)

PLASMA
leaving kidney
659 ml/minute - now contains
41 mg INULIN/100 ml *since plasma is now
diluted by the reabsorption of water only.*

By 'PLASMA CLEARANCE'
is meant the volume of
Plasma freed of the
substance by the kidney
in 1 minute.
CALCULATION:—
Plasma contained
50 mg INULIN/100 ml.
Urine contained
6000 mg INULIN/100 ml.
Urine excreted at rate of
1 ml/minute.
∴ INULIN CLEARANCE or
GLOMERULAR FILTRATION
rate $= \dfrac{6000 \times 1}{50}$
$= 120$ ml/minute

This idea of clearance can be applied to other substances naturally present such as UREA, or artificially introduced, such as DIODONE.

139

UREA "CLEARANCE"

UREA, like Inulin, is filtered from the Plasma in the Renal Corpuscle. Unlike Inulin some Urea diffuses back into the Blood Stream from the Tubules.

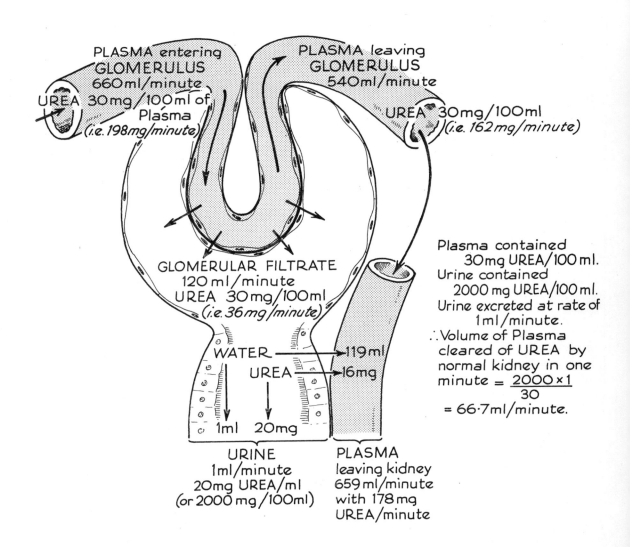

PLASMA entering
GLOMERULUS
660 ml/minute
UREA 30 mg/100 ml of Plasma
(i.e. 198 mg/minute)

PLASMA leaving
GLOMERULUS
540 ml/minute
UREA 30 mg/100 ml
(i.e. 162 mg/minute)

GLOMERULAR FILTRATE
120 ml/minute
UREA 30 mg/100 ml
(i.e. 36 mg/minute)

WATER → 119 ml
UREA → 16 mg

1 ml   20 mg

URINE
1 ml/minute
20 mg UREA/ml
(or 2000 mg/100 ml)

PLASMA
leaving kidney
659 ml/minute
with 178 mg
UREA/minute

Plasma contained
30 mg UREA/100 ml.
Urine contained
2000 mg UREA/100 ml.
Urine excreted at rate of
1 ml/minute.
∴ Volume of Plasma
cleared of UREA by
normal kidney in one
minute = $\dfrac{2000 \times 1}{30}$
= 66.7 ml/minute.

UREA clearance is used as a test of normal renal function.

DIODONE "CLEARANCE"

Certain substances are secreted by the tubules during the formation of urine. Creatinine, which is normally found in urine, is one of these.

Some foreign substances injected into the body are excreted in this way too, e.g. DIODONE *(an iodine-containing material)*.

Such substances can form the basis of tests for checking the efficiency of the Tubular epithelium in any one individual.

Some Diodone is normally *filtered;* some is *excreted* by the tubules.

Complete clearance of DIODONE from Plasma in one passage through normal kidney, gauges not only Glomerular Filtrating power but also the efficiency of the Tubular epithelium to secrete.

MAINTENANCE of ACID-BASE BALANCE

Large amounts of ACIDS are produced in cells during METABOLISM. In spite of this the pH of the blood and body fluids is kept relatively constant (7.4)

Together with the Respiratory System, the Kidneys are responsible for maintaining this equilibrium.

The ACIDS are first neutralized by BUFFERING AGENTS in the Blood Stream. They cannot be excreted in this form otherwise the body would soon lose its stocks of these ALKALIS.

The Kidney *conserves the body's reserves of Alkali* in several ways:-
1. By *Tubular Secretion of Hydrogen ions* and their *Exchange* with *Base* in the *Filtrate*.
2. By *Tubular Formation and Secretion of Ammonia* which combines with *Acids* to be excreted as Ammonium Salts

E.g.

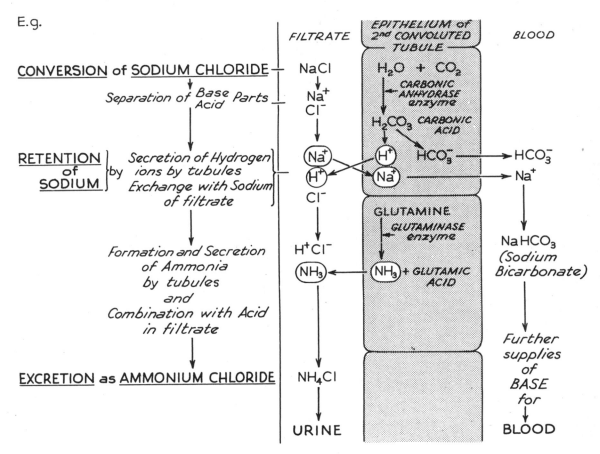

MAINTENANCE of ACID-BASE BALANCE

The chief buffering substances of the blood which appear in the filtrate are *Phosphates and Bicarbonates*. The Kidney is believed to *conserve* these as follows:-

CONVERSION of ALKALINE PHOSPHATE

Separation of Base Acid Parts

Secretion of Hydrogen ions by tubule cells

Exchange with Sodium of filtrate

RETENTION of some BASE (Sodium)

EXCRETION of ACID PHOSPHATE

FILTRATE — EPITHELIUM of 2ⁿᵈ CONVOLUTED TUBULE — BLOOD

Na_2HPO_4 — $H_2O + CO_2$ — CARBONIC ANHYDRASE enzyme — H_2CO_3 CARBONIC ACID

Na^+ — Na^+ — HPO_4^{--} — Na^+ — H^+ — HCO_3^- — HCO_3^-

H^+ — Na^+ — Na^+

Na^+ — HPO_4^{--} — $NaHCO_3$

Further supplies of Base - Sodium Bicarbonate - for

NaH_2PO_4

URINE — BLOOD

CONVERSION of SODIUM BICARBONATE

Separation of Base Acid Parts

Secretion of Hydrogen ions by tubule cells

Exchange with Sodium of filtrate

RETENTION of BASE

Carbon dioxide diffuses back to Blood

EXCRETION of WATER

FILTRATE — EPITHELIUM of 2ⁿᵈ CONVOLUTED TUBULE — BLOOD

$NaHCO_3$ — $H_2O + CO_2$ — CARBONIC ANHYDRASE enzyme

Na^+ — HCO_3^- — H_2CO_3

Na^+ — H^+ — HCO_3^- — HCO_3^-

H^+ — Na^+ — Na^+

HCO_3^- — $NaHCO_3$

H_2CO_3

CO_2 — CO_2 — CO_2

H_2O

URINE — BLOOD

REGULATION of WATER BALANCE

As the amounts of water and/or electrolytes in the body fluctuate, excretion of them is adjusted by the *Kidney* so that Body Fluids are restored to normal composition and volume.

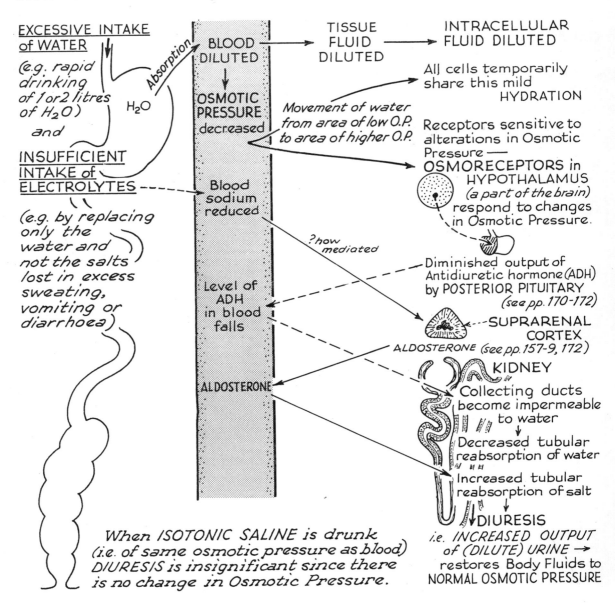

<u>EXCESSIVE INTAKE</u> <u>of WATER</u>

(e.g. rapid drinking of 1 or 2 litres of H_2O) *and*

<u>INSUFFICIENT</u> <u>INTAKE of</u> <u>ELECTROLYTES</u>

(e.g. by replacing only the water and not the salts lost in excess sweating, vomiting or diarrhoea)

H_2O

Absorption

BLOOD DILUTED

OSMOTIC PRESSURE *decreased*

Blood sodium reduced

Level of ADH in blood falls

ALDOSTERONE

TISSUE FLUID DILUTED

INTRACELLULAR FLUID DILUTED

Movement of water from area of low O.P. to area of higher O.P.

All cells temporarily share this mild HYDRATION

Receptors sensitive to alterations in Osmotic Pressure — OSMORECEPTORS in HYPOTHALAMUS *(a part of the brain)* respond to changes in Osmotic Pressure.

?how mediated

Diminished output of Antidiuretic hormone (ADH) by POSTERIOR PITUITARY *(see pp. 170-172)*

SUPRARENAL CORTEX ALDOSTERONE *(see pp. 157-9, 172)*

KIDNEY

Collecting ducts become impermeable to water

Decreased tubular reabsorption of water

Increased tubular reabsorption of salt

DIURESIS *i.e. INCREASED OUTPUT of (DILUTE) URINE →* restores Body Fluids to NORMAL OSMOTIC PRESSURE

When *ISOTONIC SALINE is drunk* *(i.e. of same osmotic pressure as blood) DIURESIS is insignificant since there is no change in Osmotic Pressure.*

144

REGULATION of WATER BALANCE

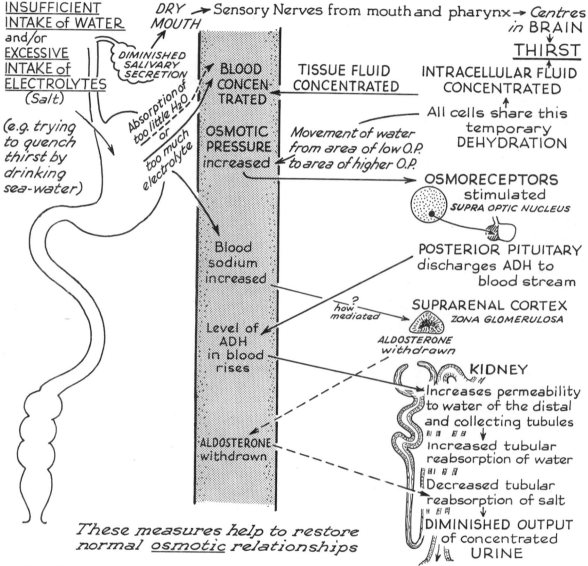

INSUFFICIENT INTAKE of WATER and/or EXCESSIVE INTAKE of ELECTROLYTES (Salt)

(e.g. trying to quench thirst by drinking sea-water)

DRY MOUTH → Sensory Nerves from mouth and pharynx → Centres in BRAIN

DIMINISHED SALIVARY SECRETION

THIRST

Absorption of too little H₂O or too much electrolyte

BLOOD CONCENTRATED

TISSUE FLUID CONCENTRATED

INTRACELLULAR FLUID CONCENTRATED

All cells share this temporary DEHYDRATION

OSMOTIC PRESSURE increased

Movement of water from area of low O.P. to area of higher O.P.

OSMORECEPTORS stimulated
SUPRA OPTIC NUCLEUS

Blood sodium increased

POSTERIOR PITUITARY discharges ADH to blood stream

how? mediated

SUPRARENAL CORTEX
ZONA GLOMERULOSA
ALDOSTERONE withdrawn

Level of ADH in blood rises

KIDNEY
Increases permeability to water of the distal and collecting tubules
Increased tubular reabsorption of water

ALDOSTERONE withdrawn

Decreased tubular reabsorption of salt

DIMINISHED OUTPUT of concentrated URINE

These measures help to restore normal osmotic relationships

In the maintenance of water balance, loss (or conservation) of water and of electrolytes are closely linked. Another example, shown on page 172, deals with ALDOSTERONE, its rôle in governing reabsorption of sodium by kidney tubules, and its tie-up with ADH in maintaining the normal volume of extracellular fluids including blood.

145

URINARY BLADDER and URETERS

A resistant, distensible *Transitional Epithelium* lines all *Urinary Passages.*

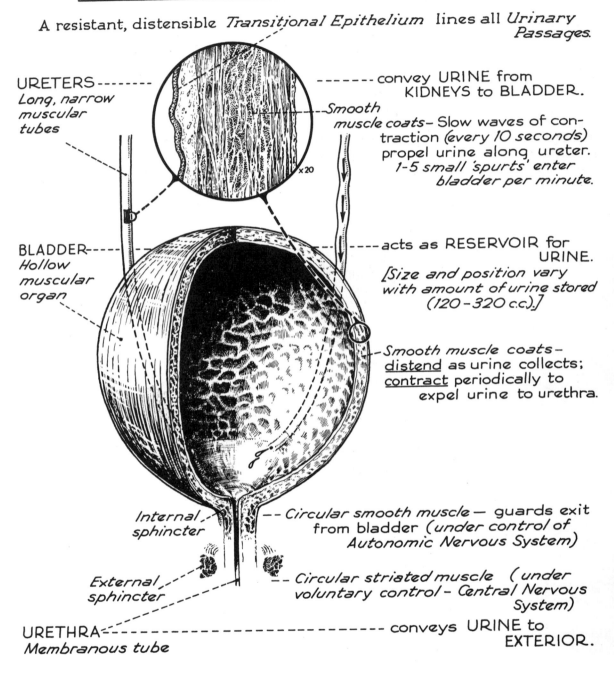

URETERS - - - - - - -
Long, narrow muscular tubes

- - - - - - - - convey URINE from KIDNEYS to BLADDER.

Smooth muscle coats– Slow waves of contraction *(every 10 seconds)* propel urine along ureter. *1-5 small 'spurts' enter bladder per minute.*

BLADDER - - - - -
Hollow muscular organ

- - - - - acts as RESERVOIR for URINE. *[Size and position vary with amount of urine stored (120-320 c.c).]*

Smooth muscle coats– <u>distend</u> as urine collects; <u>contract</u> periodically to expel urine to urethra.

Internal sphincter

- - *Circular smooth muscle*— guards exit from bladder *(under control of Autonomic Nervous System)*

External sphincter

- - *Circular striated muscle* *(under voluntary control – Central Nervous System)*

URETHRA - - - - - - -
Membranous tube

- - - - - - conveys URINE to EXTERIOR.

STORAGE and EXPULSION of URINE

URINE is formed continuously by the KIDNEYS.   It collects, drop by drop, in the URINARY BLADDER which expands to hold about 300 c.c. When the Bladder is full the desire to *void urine* is experienced.

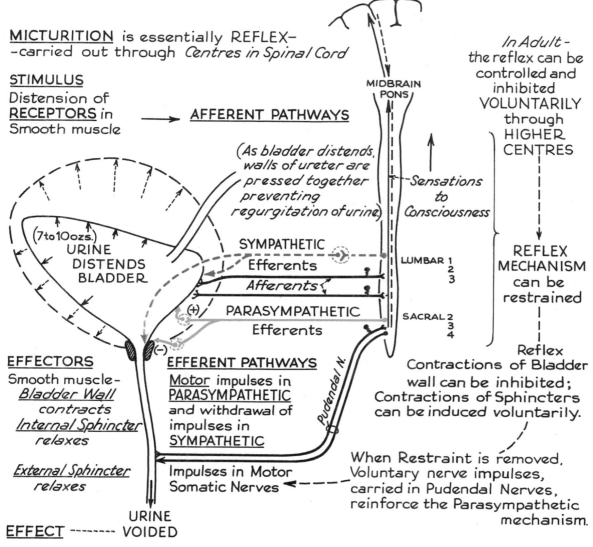

MICTURITION is essentially REFLEX—
—carried out through *Centres in Spinal Cord*

In Adult - the reflex can be controlled and inhibited VOLUNTARILY through HIGHER CENTRES

STIMULUS
Distension of **RECEPTORS** in Smooth muscle ——→ AFFERENT PATHWAYS

MIDBRAIN PONS

(As bladder distends, walls of ureter are pressed together preventing regurgitation of urine.)

–Sensations to Consciousness

(7 to 10 ozs.) URINE DISTENDS BLADDER

SYMPATHETIC
Efferents

Afferents

(+) PARASYMPATHETIC
Efferents

(−)

LUMBAR 1 2 3

SACRAL 2 3 4

REFLEX MECHANISM can be restrained

EFFECTORS
Smooth muscle-
Bladder Wall contracts
Internal Sphincter relaxes

External Sphincter relaxes

EFFERENT PATHWAYS
Motor impulses in PARASYMPATHETIC and withdrawal of impulses in SYMPATHETIC

Pudendal N.

Reflex Contractions of Bladder wall can be inhibited; Contractions of Sphincters can be induced voluntarily.

Impulses in Motor Somatic Nerves

When Restraint is removed, Voluntary nerve impulses, carried in Pudendal Nerves, reinforce the Parasympathetic mechanism.

URINE
EFFECT ------- VOIDED

When Bladder is empty and beginning to fill -
Inhibition of Parasympathetic ⎫ *Relaxation of Bladder Wall.*
Activation of Sympathetic ⎭ *Constriction of Sphincters.*

URINE

VOLUME: *In Adult* 1000 - 1500 ml/24 hours

SPECIFIC GRAVITY 1·001 - 1·040

} Vary with *Fluid Intake* and with *Fluid Output* from other routes — *Skin, Lungs, Gut.* [Volume reduced during *Sleep* and *Muscular Exercise* : Specific Gravity greater on Protein diet.]

REACTION Normally slightly acid — *Varies* with *Diet* (*pH around 6*) [acid on ordinary mixed diet: alkaline on vegetarian diet].

COLOUR

YELLOW due to UROCHROME pigment- probably from destruction of tissue protein.

More concentrated and DARKER in early morning — less water excreted at night but unchanged amounts of Urinary Solids.

ODOUR

AROMATIC when fresh→ AMMONIACAL on standing due to bacterial decomposition of UREA to AMMONIA.

COMPOSITION

Grams excreted in 24 hours

WATER 96%   1000-1500

INORGANIC SUBSTANCES

Sodium	6
Chloride	7
Calcium	0·2
Potassium	2
Phosphates	1·7
Sulphates	1·8

[These figures are approximate and vary widely in healthy individuals]

ORGANIC SUBSTANCES

Urea 2%   20-30 — derived from breakdown of *Protein* — therefore varies with Protein in Diet.

Uric Acid   0·6 — comes from *Purine* of Food and Body Tissues.

Creatinine   1·2 — from breakdown of Body Tissues; uninfluenced by amount of dietary Protein.

Ammonia   0·5-0·9 — formed in Kidney from *Glutamine* brought to it by Blood Stream; varies with amounts of acid substances requiring neutralization in the Kidney.

[In the Newborn, Volume and Specific Gravity are low and Composition varies.]

148

CHAPTER 7

ENDOCRINE SYSTEM

ENDOCRINE SYSTEM

The <u>DUCTLESS GLANDS</u> produce HORMONES ("chemical messengers") which they pass into the BLOOD STREAM for GENERAL CIRCULATION to excite or inhibit activity of other organs or tissues.

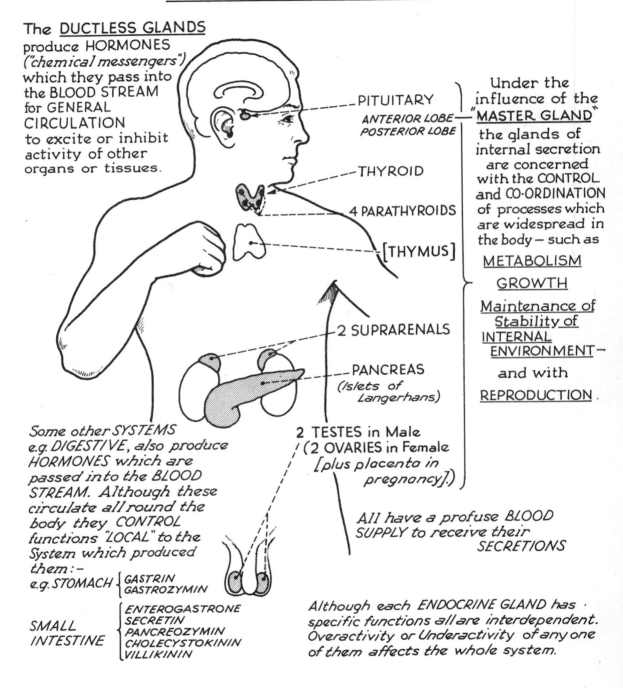

PITUITARY
ANTERIOR LOBE
POSTERIOR LOBE

THYROID

4 PARATHYROIDS

[THYMUS]

2 SUPRARENALS

PANCREAS
(Islets of Langerhans)

2 TESTES in Male
(2 OVARIES in Female
[plus placenta in pregnancy].)

Under the influence of the "<u>MASTER GLAND</u>" the glands of internal secretion are concerned with the CONTROL and CO-ORDINATION of processes which are widespread in the body — such as

<u>METABOLISM</u>

<u>GROWTH</u>

<u>Maintenance of Stability of INTERNAL ENVIRONMENT</u> — and with

<u>REPRODUCTION</u>.

All have a profuse BLOOD SUPPLY to receive their SECRETIONS

Some other SYSTEMS e.g. DIGESTIVE, also produce HORMONES which are passed into the BLOOD STREAM. Although these circulate all round the body they CONTROL functions "LOCAL" to the System which produced them :-
e.g. STOMACH { *GASTRIN*
GASTROZYMIN

SMALL INTESTINE { *ENTEROGASTRONE*
SECRETIN
PANCREOZYMIN
CHOLECYSTOKININ
VILLIKININ

Although each ENDOCRINE GLAND has specific functions all are interdependent. Overactivity or Underactivity of any one of them affects the whole system.

THYROID

STRUCTURE:

2 LOBES (joined by ISTHMUS) composed of FOLLICLES lined by CUBICAL EPITHELIUM

lie in front of TRACHEA

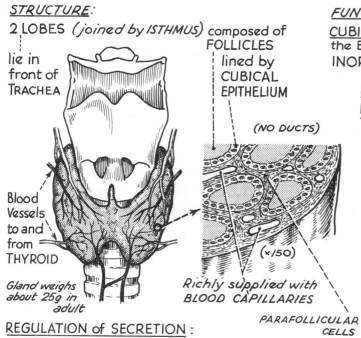

Blood Vessels to and from THYROID

Gland weighs about 25g in adult

(NO DUCTS)

(×150)

Richly supplied with BLOOD CAPILLARIES

PARAFOLLICULAR CELLS

FUNCTION:

CUBICAL EPITHELIUM extracts from the Blood stream (and concentrates) INORGANIC IODIDE

IODINE links with TYROSINE } MONOIODOTYROSINE DIIODO-TYROSINE

BLOOD

TRIIODO-THYRONINE
TETRAIODO-THYRONINE (Thyroxine)

Stored in colloid linked with protein as THYROGLOBULIN | when required a protein-splitting enzyme releases active agents

REGULATION of SECRETION:

HYPOTHALAMUS

FALL in blood THYROID HORMONES

promotes production of THYROID-STIMULATING HORMONE(TSH) in ANTERIOR PITUITARY

stimulates production and release of

THYROTROPHIN-RELEASING FACTOR (TRF)

RISE in blood THYROID HORMONES

depresses production of THYROID-STIMULATING HORMONE (TSH or THYROTROPHIN) in ANTERIOR PITUITARY

THYROID HORMONES

THYROID GLAND

BODY TISSUES

The ACTIVE HORMONES are released to the GENERAL CIRCULATION for distribution by the blood stream to all BODY TISSUES where they act as CATALYSTS hastening OXIDATION processes in the TISSUE CELLS. i.e. They probably regulate the various ENZYMES which control ENERGY METABOLISM and thus influence all METABOLIC PROCESSES including GROWTH.

[PARAFOLLICULAR Cells may secrete THYROCALCITONIN – which LOWERS a high BLOOD CALCIUM probably by suppressing calcium mobilization from bone.]

UNDERACTIVITY of THYROID

If the THYROID shows *atrophy* of its secretory cells or is inadequately stimulated by the Anterior Pituitary:-

FALL (or absence) of THYROID HORMONES in blood

LESS (or no) THYROID HORMONES enter blood to depress activity of ANTERIOR PITUITARY

Less functional tissue to respond to THYROID STIMULATING HORMONE

Body Tissues

x200

Insufficient HORMONAL SECRETION released to Blood Stream.
TISSUE OXIDATIONS are depressed, *i.e.* Rate at which cells use energy is reduced.
The Basal Metabolic Rate falls.
Less Heat is produced.
Body Temperature falls (and person feels COLD).
Energy stores increase (e.g. GLYCOGEN and FAT).

In the <u>ADULT</u>

MYXOEDEMA

<u>SLOWING UP OF ALL BODILY PROCESSES</u>

Blood cholesterol increases.
{ Appetite is reduced; Weight increases. Gut movements are sluggish → Constipation. Heart and Respiratory Rates and Blood Pressure reduced. Thought processes slow down → Lethargy; Apathy. *SKIN* — Thick, leathery, puffy. *HAIR* - Brittle, sparse, dry. }

In the <u>CHILD</u> — congenital absence of the gland → <u>CRETIN</u>

DWARFING

{ FAILURE of SKELETAL SEXUAL MENTAL } GROWTH and DEVELOPMENT

All "milestones" of babyhood are delayed.

THYROID EXTRACT (taken by mouth) restores individuals to normal.

152

OVERACTIVITY of THYROID

If an *enlarged* THYROID shows increased activity of its secretory cells:-

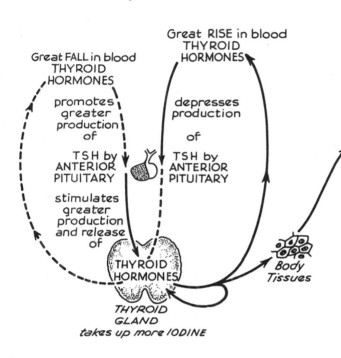

Great RISE in blood THYROID HORMONES

Great FALL in blood THYROID HORMONES

promotes greater production of TSH by ANTERIOR PITUITARY

stimulates greater production and release of

depresses production of TSH by ANTERIOR PITUITARY

THYROID HORMONES

THYROID GLAND
takes up more IODINE

Body Tissues

EXCESS THYROID HORMONES are distributed by blood stream to the Tissues of the Body. ⟶
SPEED up OXIDATIONS in the cells, i.e rate at which all cells use ENERGY.

The Basal Metabolic Rate is raised. As a by-product of this increased cellular activity more heat is produced ⟶ rise in Body Temperature (*person feels* WARM). ↓
Profuse Sweating.
Energy stores of body (i.e.GLYCOGEN and FAT) are depleted.

SPEEDING UP
OF ALL
BODILY
PROCESSES

⎧ Appetite increases but weight falls.
Movements of digestive tract are increased ⟶ Diarrhoea.
Heart and Respiratory Rates rise.
Blood Pressure is raised.
Muscular tremor and nervousness are marked.
Person becomes excitable and
apprehensive. ⎩

*[*EXOPHTHALMOS (*protrusion of eyeballs*) may be due to an excess of some Pituitary Hormone. It is not due to an excess of Thyroid Hormones.*]*

Surgical removal of part of the overactive gland reduces the Thyroid activity.

PARATHYROIDS

FOUR small glands composed of *cords of cells* which secrete Parathyroid
Hormone – **PARATHORMONE**
or **PTH**

CAPILLARIES

GENERAL CIRCULATION

to ALL TISSUES of the
body

But not all tissues are
sensitive to it.

It plays an important
rôle in CALCIUM and
PHOSPHATE
METABOLISM

PARATHYROID
GLANDS

Situated
behind
THYROID

OESOPHAGUS

TRACHEA

Each weighs from
20-50mg in Adult

x 200

Function of
EOSINOPHIL
cells is
unknown

PARATHORMONE acts primarily on <u>KIDNEY TUBULES</u>, <u>BONE</u> and on <u>GUT</u> to maintain
BLOOD CALCIUM level at 11mg/100ml PLASMA (*necessary for normal neuromuscular excitability*).

Alterations in concentration of CALCIUM ions in the blood are said to control Parathyroid activity. A rise in blood Calcium depresses Parathyroid secretion. A fall in Calcium increases Parathyroid secretion. [Note: This response to the level of plasma Ca differs from thyrocalcitonin's – see 'thyroid'.]

154

UNDERACTIVITY of PARATHYROIDS

Atrophy or removal of Parathyroid tissue causes a fall in BLOOD CALCIUM level and increased excitability of Neuromuscular tissue.   This leads to severe convulsive disorder — TETANY.

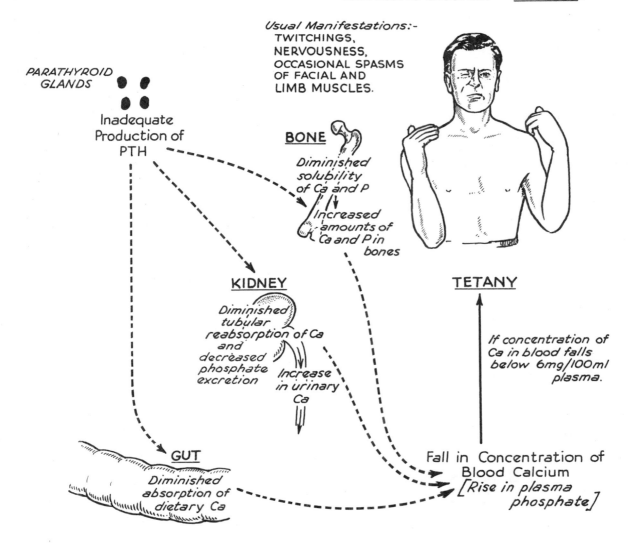

Usual Manifestations:- TWITCHINGS, NERVOUSNESS, OCCASIONAL SPASMS OF FACIAL AND LIMB MUSCLES.

PARATHYROID GLANDS

Inadequate Production of PTH

BONE

Diminished solubility of Ca and P

Increased amounts of Ca and P in bones

KIDNEY

Diminished tubular reabsorption of Ca and decreased phosphate excretion

Increase in urinary Ca

TETANY

If concentration of Ca in blood falls below 6mg/100ml plasma.

GUT

Diminished absorption of dietary Ca

Fall in Concentration of Blood Calcium [*Rise in plasma phosphate*]

[*Note the inverse relationship between plasma calcium and inorganic phosphate*]

Symptoms are relieved by injection of Calcium or of Extract of Parathyroid.

OVERACTIVITY of PARATHYROIDS

Overactivity of the Parathyroids *(due often to tumour)* leads to rise in BLOOD CALCIUM level and eventually to <u>OSTEITIS FIBROSA CYSTICA</u>.

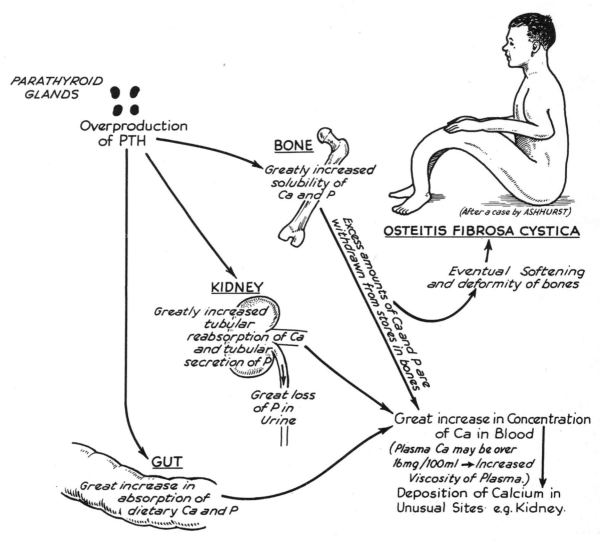

PARATHYROID GLANDS

Overproduction of PTH

BONE
Greatly increased solubility of Ca and P

Excess amounts of Ca and P are withdrawn from stores in bones

(After a case by ASHHURST)

OSTEITIS FIBROSA CYSTICA

Eventual Softening and deformity of bones

<u>KIDNEY</u>
Greatly increased tubular reabsorption of Ca and tubular secretion of P

Great loss of P in Urine

<u>GUT</u>
Great increase in absorption of dietary Ca and P

Great increase in Concentration of Ca in Blood *(Plasma Ca may be over 16mg/100ml → Increased Viscosity of Plasma.)* Deposition of Calcium in Unusual Sites· e.g. Kidney.

The increased level of blood calcium eventually leads to excessive loss of CALCIUM in URINE and also of WATER since the salts are excreted in solution.

Excision of the overactive Parathyroid tissue abolishes syndrome.

SUPRARENAL GLANDS

There are TWO Suprarenal Glands. They lie close to the kidneys.

Each has an outer CORTEX and an inner MEDULLA

RIGHT KIDNEY

LEFT KIDNEY

CORTEX

CAPSULE
ZONA GLOMERULOSA
ZONA FASCICULATA
ZONA RETICULARIS
MEDULLA

× 80

The Suprarenal Cortex secretes groups of steroid HORMONES into the BLOOD STREAM. These travel to all tissues of the body.
The Suprarenal Steroids share the same functions but to varying degrees:—

1. "MINERALOCORTICOIDS"
(e.g. Deoxycorticosterone and Aldosterone) have chief action on <u>KIDNEY TUBULES</u>

KIDNEY

They promote <u>RETENTION</u> of <u>SODIUM</u> (*with water*)
Stimulate <u>EXCRETION</u> of <u>POTASSIUM</u>

2. "GLUCOCORTICOIDS"
(e.g. Hydrocortisone, Cortisone, Corticosterone) – on <u>LIVER</u>

LIVER

Stimulate <u>FORMATION</u> of <u>SUGAR from PROTEIN</u>
(*This gives increased BLOOD SUGAR and leads to increased GLYCOGEN stores*)
1 and 2 have Anti-inflammatory and Anti-allergic properties.

3. <u>SEX HORMONES</u>
allied to ANDROGENS of TESTIS and OESTROGEN and PROGESTERONE of OVARY.
[Both groups are present in both sexes]
These sex hormones may influence development of SEX FUNCTIONS and of Secondary Sex characteristics

OVARY

TESTIS

SALT and WATER metabolism [Electrolyte balance]

CARBOHYDRATE metabolism

SEX FUNCTIONS

Secretion is under control of the ANTERIOR PITUITARY ADRENO-CORTICOTROPHIC HORMONE (*CORTICOTROPHIN, ACTH*)

STRESS acts via HYPOTHALAMUS

CORTICOTROPHIN release factor

FALL in blood CORTICOIDS

promotes production of CORTICOTROPHIN stimulates production and release of CORTICOIDS

RISE in blood CORTICOIDS

depresses production of CORTICOTROPHIN

SUPRARENAL GLAND

CORTICOIDS

This reciprocal relationship between A.P. and S.R. Cortex leads to balanced effects on →

The most powerful Salt retaining Hormone – ALDOSTERONE – is outwith Pituitary control. Its production in the Zona Glomerulosa is apparently stimulated by sodium deprivation (p.144) and by reduction in volume of extracellular fluid (p.172).

The Suprarenal Cortex is <u>essential to life</u> and plays an important rôle in states of stress.

UNDERACTIVITY of SUPRARENAL CORTEX

Atrophy of Suprarenal Cortex *(occasionally occurs with destructive disease*
            gives   *of the gland, e.g. Tuberculosis).*
Inadequate Production of CORTICOIDS :-

"MINERALOCORTICOIDS"   ADDISON'S DISEASE

INADEQUATE REABSORPTION of Sodium and Water and Excessive Reabsorption of Potassium

Blood Sodium level falls → BODY FLUIDS show grave ELECTROLYTIC imbalance

Potassium level rises → Potassium escapes into TISSUE FLUIDS

(Increase in Blood Urea eventually)

Excessive loss of Sodium and Water in Urine

Potassium level falls

NORMAL OSMOTIC RELATIONSHIPS are not maintained

Blood Volume falls

Blood Pressure falls

Inadequate Production of "GLUCOCORTICOIDS"

GLUCOSE formation from PROTEIN is depressed - - - - → Blood Sugar level falls

Patients show —{ GREAT MUSCULAR WEAKNESS and WASTING, GREAT LOSS of WEIGHT, GASTRO-INTESTINAL UPSETS *(vomiting, diarrhoea)*, ANAEMIA, PIGMENTATION of exposed and pressure areas of SKIN.

If untreated - eventually fatal.

Administration of Suprarenal Hormones and Sodium Chloride restores individual to normal.

158

OVERACTIVITY of SUPRARENAL CORTEX

Overactivity or tumour of Suprarenal Cortex
          may give
Overproduction of any or all of the CORTICOIDS:-

"MINERALOCORTICOIDS"
          e.g. ALDOSTERONE ⟶ ALDOSTERONISM

EXCESSIVE REABSORPTION of Sodium and Water

Blood Sodium level rises ⟶

Diminished loss of Sodium and Water in Urine

Upset in normal OSMOTIC PRESSURE RELATIONSHIPS ⟶ Excessive Salt and Water in TISSUES

↓

OEDEMA

Overproduction of "GLUCOCORTICOIDS" (chiefly of Hydrocortisone) ⟶ CUSHING'S SYNDROME

Sometimes occur together

Excess GLUCOSE formation from PROTEIN ⟶ High Blood Sugar ⟶ Obesity especially of face and trunk.

Glucose in Urine

Overproduction of Suprarenal SEX HORMONES ⟶ ADRENOGENITAL SYNDROME
(Frequently excess production of Androgens)

In CHILDREN precocious development of Secondary Sex characteristics

Virilism in 9 year old girl

Virilism in ADULT woman

Muscular hypertrophy in 6 year old boy 'Pocket Hercules'

(After a case by KEPLER et al.)

(After a case by GUTHRIE)

Administration of CORTISONE depresses Pituitary secretion of ACTH ⟶ inhibits production of the abnormal steroids.
   Removal of the over-secreting tissue or tumour restores individual.

SUPRARENAL MEDULLA

The Suprarenal Medulla arises from the same primitive tissue as the Ganglion cells of the Sympathetic Nervous System.

The cells contain granules which stain brown with Salts of CHROMIC acid. These CHROMAFFIN granules are the precursors of ⋯⋯ ADRENALINE and

CENTRAL VEIN

×200

Richly supplied by a plexus of Sympathetic Nerve Fibres

A second cell type may secrete — NORADRENALINE

released to
CAPILLARIES
↓
GENERAL CIRCULATION
↓
BODY TISSUES

Secretion of the SUPRARENAL MEDULLA is under control of the SYMPATHETIC NERVOUS SYSTEM

In times of STRESS or FEAR

Frontal Lobes stimulate one area of HYPOTHALAMUS

Centre in MEDULLA OBLONGATA

sends impulses along Sympathetic Nerves (Greater Splanchnic) which cause liberation of Adrenaline by Suprarenal Medullae into Blood Stream

O₂ lack in Blood Stream has a direct effect on Suprarenal Medullae

O₂

Adrenaline reinforces action of Sympathetic Nervous System in preparing the various SYSTEMS of the BODY to react efficiently in emergencies and stress.

There is some evidence to suggest that Adrenaline and Noradrenaline are released separately. e.g. Stimulation of another part of the Hypothalamus apparently leads to release of Noradrenaline into Blood Stream
↓
general vasoconstriction (except of coronaries)
↓
Rise in Blood Pressure.

160

ADRENALINE

Under quiet resting conditions the blood contains very little adrenaline. During excitement or circumstances which demand special efforts adrenaline is released into the blood stream, and is responsible for the following actions summed up as the "*FIGHT or FLIGHT*" FUNCTION of the Suprarenal medullae.

It _Constricts_ Smooth Muscle of Skin → Hairs 'stand on end'; 'Gooseflesh'.

Dilates Pupil of Eye to admit more light.

Constricts Smooth Muscle of Abdominal Blood Vessels and Cutaneous Blood Vessels → Pallor with Fright.

Dilates Smooth Muscle in Blood Vessels of Heart (Coronaries) and of Skeletal Muscles, i.e. better supply to organs requiring it in emergency.

Increases Heart Rate and Cardiac Output. Blood Pressure raised.

Relaxes Smooth Muscle in Wall of Bronchioles → better supply of air to alveoli.

Stimulates Respiration.

Inhibits Movements of Digestive Tract.

Contracts Sphincters of Gut.

Inhibits Wall of Urinary Bladder.

Contracts Ureters and Sphincter of Urinary Bladder.

Mobilizes Muscle and Liver Glycogen → increase in Blood Sugar.

Stimulates Metabolism.

Exerts favourable effect on contracting Skeletal Muscle → Fatigues less readily.

Increases Coagulability of Blood.

Most of these effects can also be produced by stimulating Sympathetic Nerve Fibres.

The Suprarenal Medullae are not essential to life — but without them the body is less able to face emergencies and conditions of stress.

DEVELOPMENT of PITUITARY

The Pituitary Gland consists of ANTERIOR and POSTERIOR parts which differ in ORIGIN, STRUCTURE and FUNCTION.

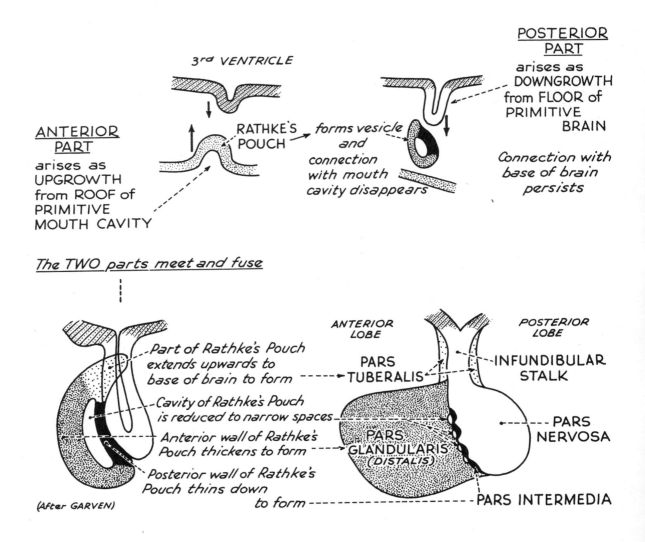

3rd VENTRICLE

POSTERIOR
PART
arises as
DOWNGROWTH
from FLOOR of
PRIMITIVE
BRAIN

Connection with
base of brain
persists

ANTERIOR
PART
arises as
UPGROWTH
from ROOF of
PRIMITIVE
MOUTH CAVITY

RATHKE'S
POUCH

forms vesicle
and
connection
with mouth
cavity disappears

The TWO parts meet and fuse

ANTERIOR
LOBE

POSTERIOR
LOBE

Part of Rathke's Pouch
extends upwards to
base of brain to form

PARS
TUBERALIS

INFUNDIBULAR
STALK

Cavity of Rathke's Pouch
is reduced to narrow spaces

Anterior wall of Rathke's
Pouch thickens to form

PARS
GLANDULARIS
(DISTALIS)

PARS
NERVOSA

Posterior wall of Rathke's
Pouch thins down
to form

PARS INTERMEDIA

(After GARVEN)

The ADULT PITUITARY (HYPOPHYSIS) is a small oval gland which lies in
the SELLA TURCICA — a small cavity in the bone at the base of the skull.

ANTERIOR PITUITARY

This is the MASTER GLAND of the ENDOCRINE SYSTEM. It regulates the activity of the other Endocrine Glands, including the GONADS, and influences ALL METABOLIC PROCESSES including GROWTH.

CORTEX

STRESS
NERVOUS REFLEXES

HYPOTHALAMUS

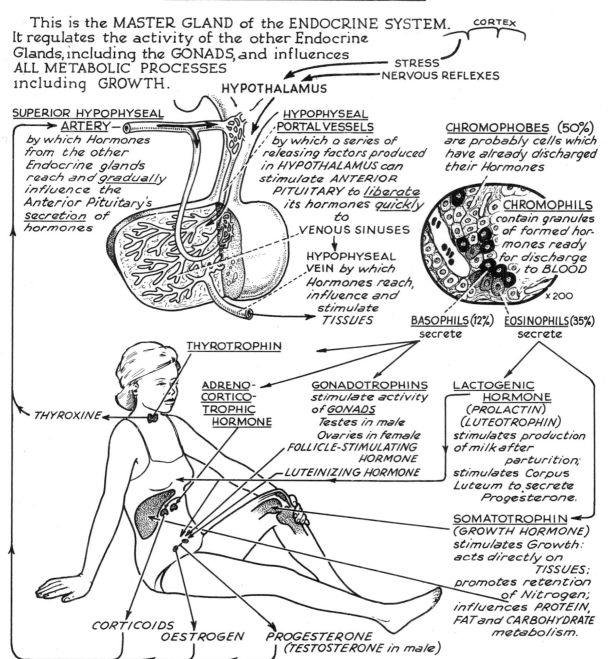

SUPERIOR HYPOPHYSEAL ARTERY — by which Hormones from the other Endocrine glands reach and *gradually* influence the Anterior Pituitary's *secretion* of hormones

HYPOPHYSEAL PORTAL VESSELS by which a series of releasing factors produced in HYPOTHALAMUS can stimulate ANTERIOR PITUITARY to *liberate* its hormones *quickly* to VENOUS SINUSES

HYPOPHYSEAL VEIN by which Hormones reach, influence and stimulate TISSUES

CHROMOPHOBES (50%) are probably cells which have already discharged their Hormones

CHROMOPHILS contain granules of formed hormones ready for discharge to BLOOD

× 200

BASOPHILS (12%) secrete

EOSINOPHILS (35%) secrete

THYROTROPHIN

THYROXINE

ADRENO-CORTICO-TROPHIC HORMONE

GONADOTROPHINS *stimulate activity of GONADS* Testes in male Ovaries in female FOLLICLE-STIMULATING HORMONE LUTEINIZING HORMONE

LACTOGENIC HORMONE (PROLACTIN) (LUTEOTROPHIN) *stimulates production of milk after parturition; stimulates Corpus Luteum to secrete Progesterone.*

SOMATOTROPHIN (GROWTH HORMONE) *stimulates Growth: acts directly on TISSUES; promotes retention of Nitrogen; influences PROTEIN, FAT and CARBOHYDRATE metabolism.*

CORTICOIDS

OESTROGEN

PROGESTERONE (TESTOSTERONE in male)

163

UNDERACTIVITY of ANTERIOR PITUITARY

Deficiency or absence
of **EOSINOPHIL** cells

Underproduction of
GROWTH Hormone
(*Somatotrophin*)

LORAIN DWARF

Delayed Skeletal
Growth and
Retarded Sexual
Development
but alert, intelligent,
well proportioned
child.

Destructive disease of part of Anterior
Pituitary (usually with damage to
Posterior Pituitary and/or Hypothalamus)

Underproduction of GROWTH and other
ENDOCRINE-TROPHIC Hormones

FRÖHLICH'S DWARF

Stunting of Growth,
Obesity (*Large
appetite for sugar*);
Arrested Sexual
Development;
Lethargic;
Somnolent;
Mentally
Subnormal.

Atrophy of
other
Endocrine
glands

Signs of
deficiency
of their
hormones.

AGE 13

NORMAL CHILD
AGE 13

AGE 13

Extracts of human GROWTH
hormone restore growth
and development pattern
to normal.

A similar condition occurs in ADULT
without dwarfing but with suppression
of sex functions and regression of
secondary sex characteristics.
Extracts of GROWTH and GONADOTROPHIC
hormones aid in restoring patient to normal.

OVERACTIVITY of PITUITARY EOSINOPHIL CELLS

Functional overactivity (or tumour) chiefly of the EOSINOPHIL cells of the Anterior Pituitary leads to ⟶ GIANTISM in the CHILD: ACROMEGALY in the ADULT.

Overproduction of GROWTH Hormone

General Circulation

Increases NITROGEN retention. Influences Protein, Carbohydrate and Fat metabolism of ALL CELLS of the body.

Overgrowth of all Body Tissues

Onset before bony epiphyses have closed at puberty

Onset after puberty

Bones thicken, especially of FACE, JAW, NOSE, HANDS and FEET ⟶

Long bones grow in length (Height 7-8 feet). Overgrowth of MUSCLES

Overgrowth of SOFT TISSUES and INTERNAL ORGANS (e.g. Heart, Spleen, Stomach, etc.)

⟶ Coarse thick SKIN

[These patients frequently show a raised Basal Metabolic rate; a high Blood Sugar level and the presence of Sugar in the urine. Other features of this condition are due often to pressure of tumour on surrounding brain tissue or sometimes to overproduction of other Anterior Pituitary Hormones.]

NORMAL CHILD
AGE 13

AGE 13

Destruction of the overactive tissue – usually by RADIUM therapy – prevents progression of the condition.

OVERACTIVITY of PITUITARY BASOPHIL CELLS

Overactivity *(often due to Tumour)* of the Basophil cells of the Anterior Pituitary gives

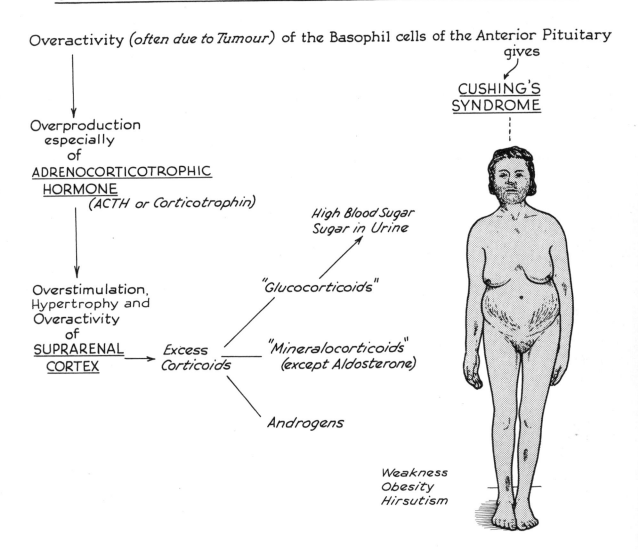

CUSHING'S SYNDROME

Overproduction especially of ADRENOCORTICOTROPHIC HORMONE *(ACTH or Corticotrophin)*

Overstimulation, Hypertrophy and Overactivity of SUPRARENAL CORTEX → Excess Corticoids

"Glucocorticoids" → High Blood Sugar Sugar in Urine

"Mineralocorticoids" (except Aldosterone)

Androgens

Weakness Obesity Hirsutism

This condition is usually indistinguishable clinically from that seen in primary overactivity or tumour of the Suprarenal Cortex itself. The syndrome is here shown in the adult woman.

Overproduction of THYROTROPHIN ⟶ *Overactivity* of THYROID *gland.*

PANHYPOPITUITARISM

Complete Atrophy (or insufficiency) of all secreting cells of Anterior Pituitary in Adult — <u>SIMMOND'S DISEASE</u>

FAILURE to PRODUCE ANY HORMONES ⟶

LACK of GROWTH HORMONE
↘ _Grave upset in TISSUE metabolism_ ⟶

LACK of GONADOTROPHINS ⟶

LACK of ENDOCRINE-TROPHIC HORMONES ⟶

<u>APPEARANCE of PREMATURE SENILITY</u>

Features usually associated with very OLD AGE

<u>HAIR</u> _grey, sparse: loss of body hair._
<u>SKIN</u> _dry, sallow, wrinkled._
<u>BODY</u> _emaciated (great loss of weight)_
<u>BONES</u> _frail_

<u>SEX ORGANS</u> _atrophy._
Menstruation ceases.
Reproductive cycle stops.
Secondary sex characteristics gradually regress.

<u>ALL ENDOCRINES</u>
atrophy and show depressed secretion of their Hormones

(Mental changes supervene)

⟶ Basal metabolism depressed
↓
Body temperature depressed,
Heart rate low.
Blood pressure low.
Blood sugar low.
Electrolytic upset.

[_This patient is only 42 years of age_]
After ZONDEK, _Diseases of the Endocrine Glands._

Extracts of Anterior Pituitary may relieve the condition but rarely succeed in completely restoring the patient to normal.

POSTERIOR PITUITARY

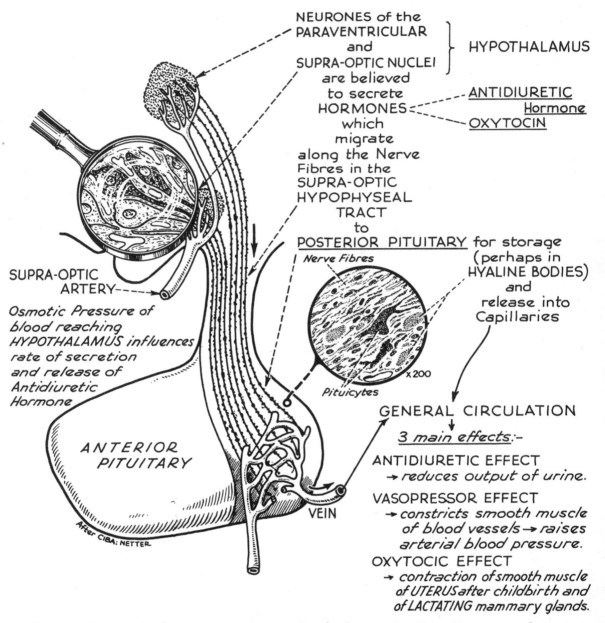

NEURONES of the
PARAVENTRICULAR
and
SUPRA-OPTIC NUCLEI — HYPOTHALAMUS
are believed
to secrete
HORMONES — ANTIDIURETIC Hormone
which — OXYTOCIN
migrate
along the Nerve
Fibres in the
SUPRA-OPTIC
HYPOPHYSEAL
TRACT
to
POSTERIOR PITUITARY for storage
(perhaps in
HYALINE BODIES)
and
release into
Capillaries

Nerve Fibres

SUPRA-OPTIC
ARTERY—

*Osmotic Pressure of
blood reaching
HYPOTHALAMUS influences
rate of secretion
and release of
Antidiuretic
Hormone*

x200

Pituicytes

*ANTERIOR
PITUITARY*

After CIBA: NETTER

VEIN

GENERAL CIRCULATION

3 main effects:-

ANTIDIURETIC EFFECT
→ *reduces output of urine.*

VASOPRESSOR EFFECT
→ *constricts smooth muscle
of blood vessels → raises
arterial blood pressure.*

OXYTOCIC EFFECT
→ *contraction of smooth muscle
of UTERUS after childbirth and
of LACTATING mammary glands.*

*There are no obvious secreting cells in Posterior Pituitary such as are
seen in other ENDOCRINE GLANDS.*

OXYTOCIN

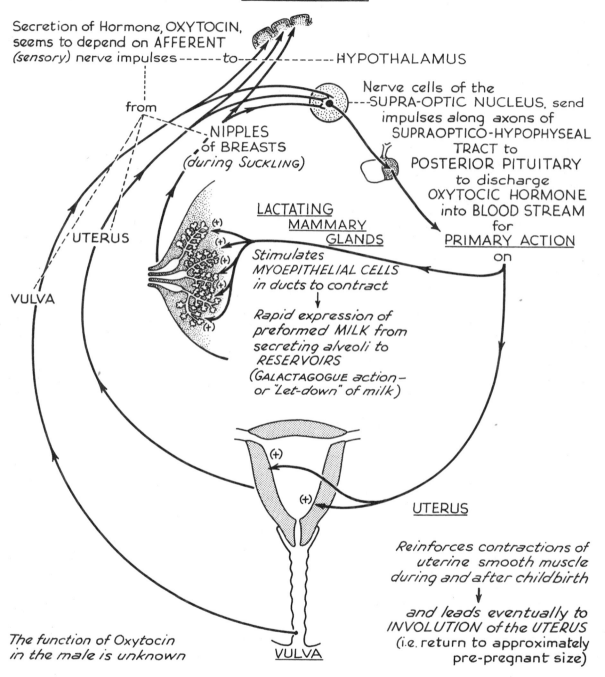

Secretion of Hormone, OXYTOCIN, seems to depend on AFFERENT (sensory) nerve impulses - - - - - to - - - - - - - - HYPOTHALAMUS

from

NIPPLES of BREASTS (during Suckling)

Nerve cells of the - - SUPRA-OPTIC NUCLEUS, send impulses along axons of SUPRAOPTICO-HYPOPHYSEAL TRACT to POSTERIOR PITUITARY to discharge OXYTOCIC HORMONE into BLOOD STREAM for PRIMARY ACTION on

UTERUS

VULVA

LACTATING MAMMARY GLANDS

Stimulates MYOEPITHELIAL CELLS in ducts to contract

Rapid expression of preformed MILK from secreting alveoli to RESERVOIRS

(GALACTAGOGUE action – or "Let-down" of milk)

UTERUS

Reinforces contractions of uterine smooth muscle during and after childbirth

and leads eventually to INVOLUTION of the UTERUS (i.e. return to approximately pre-pregnant size)

The function of Oxytocin in the male is unknown

VULVA

ANTIDIURETIC HORMONE

INCREASED Concentration of PLASMA SALT
due to
LACK of Dietary WATER or to
LOSS of BODY WATER from
      SWEAT GLANDS *(in sweat)*
      LUNGS *(in expired air)*
      GUT *(in faeces)*
      or to
EXCESS DIETARY SALT
results in

INCREASED OSMOTIC PRESSURE
of BLOOD ------- stimulates
OSMORECEPTORS

STIMULI from
EXTERNAL
ENVIRONMENT
(e.g. painful
stimuli)
and
EMOTIONAL
STRESS

NERVOUS
REFLEXES

Stimulate

SUPRAOPTIC NUCLEUS

Nerve cells transmit
impulses along axons of
SUPRAOPTICO-HYPOPHYSEAL
    TRACT to
POSTERIOR PITUITARY
to cause discharge of
ANTIDIURETIC HORMONE (ADH)
into BLOOD STREAM for
DIRECT ACTION on CELLS
LINING DISTAL and COLLECT-
ING TUBULES of KIDNEY
        NEPHRON
ADH increases their
  permeability to water
      ↓
Augments reabsorption of
WATER from GLOMERULAR
    FILTRATE in
DISTAL (2ND) CONVOLUTED
TUBULE and in COLLECTING
       TUBULES
      ↓
Reduces output of URINE to
BALANCE and RESTORE
OSMOTIC PRESSURE
RELATIONSHIPS to normal.

[Diminished Concentration
of Plasma Salt due to
e.g. Excess Intake of Water

Dilution of Blood Stream
    ↓
Diminished Osmotic
* Pressure*
    ↓
Fall in Activity of the
* Osmoreceptors*
    ↓
Fall in Output of ADH
    ↓
Diminished Reabsorption
* of Water*
    ↓
Increased Output
of Dilute Urine
    ↓
Restores OSMOTIC
* PRESSURE*
* RELATIONSHIPS*
* to normal]*

maintain
FLUID EQUILIBRIUM

(This hormone also has vasoconstrictor effects → increased Blood Pressure)

UNDERACTIVITY of POSTERIOR PITUITARY

Damage, by INJURY or DISEASE, to

X----- HYPOTHALAMUS

or to

X----- SUPRAOPTICO-HYPOPHYSEAL TRACT

or to

X----- POSTERIOR PITUITARY   causes   DIABETES INSIPIDUS
characterized by
EXCESSIVE THIRST and
EXCESSIVE PRODUCTION
of DILUTE URINE

Absence from
Blood Stream of
ANTIDIURETIC HORMONE
(ADH)

DIMINISHED
REABSORPTION of WATER   from   Normal GLOMERULAR
FILTRATE-of over 100 litres/
day

[OBLIGATORY REABSORPTION   of   about 70 to 80 litres of
from   glomerular filtrate water are
PROXIMAL (1ST) CONVOLUTED   _   reabsorbed outwith
TUBULE   influence of ADH.]

FACULTATIVE REABSORPTION   of   about 20 to 30 litres per
from   day
DISTAL (2ND) CONVOLUTED
TUBULE and COLLECTING   _   normally under ADH
DUCT does not occur   control.
(Cells lining collecting duct
remain impermeable to water)

Increased ELIMINATION
of WATER

URINARY VOLUME rises ————   up to 20 or 30 litres of PALE
DILUTE URINE excreted/day
(SPECIFIC GRAVITY 1.001 to 1.002)
instead of normal 1-1½ litres
straw coloured more
concentrated fluid
(S.G. 1.020 – 1.032)

Small amounts of Posterior Pituitary extract absorbed from under the tongue or given by subcutaneous injection reduce elimination of water to normal.

ALDOSTERONE and ANTIDIURETIC HORMONE (ADH) and the MAINTENANCE of BLOOD VOLUME

CIRCULATION

A reduction (↓) in BLOOD VOLUME (*as after haemorrhage*) leads to temporary ↓CARDIAC OUTPUT

↓ B.P.

ANGIOTENSINOGEN (*part of α_2 globulin*)

→RENIN (*a protein-splitting enzyme*)

ANGIOTENSIN (*powerful vaso-constrictor*)

↑ B.P.

↑ALDOSTERONE

↑Na⁺

↑O.P.

↑ADH

↑H₂O (*this together with ↑Na⁺*)

↑BLOOD VOLUME (*i.e. restores to normal*)

TISSUE FLUIDS

[*or reduction in total volume of EXTRACELLULAR FLUID (such as may result from loss of ISOTONIC SECRETIONS from GUT).*]

VOMITING

DIARRHOEA

ENDOCRINES

SUPRARENAL CORTEX
Zona Glomerulosa secretes

ALDOSTERONE

HYPOTHALAMUS OSMORECEPTORS *secrete.....and* POSTERIOR PITUITARY *releases*

↑ ADH

KIDNEY

↓PULSE PRESSURE in AFFERENT ARTERIOLE (*i.e. less stretch applied to walls*)

JUXTAGLOMERULAR APPARATUS *releases*

RENIN

DISTAL TUBULE *promotes* Na⁺ retention↑

DISTAL and COLLECTING TUBULES *rendered permeable to water* H₂O reabsorption↑

Aldosterone and ADH co-operate in other ways to maintain salt and water balance e.g. where osmotic pressure relationships (and not primarily volumes) are disturbed (see pp. 144, 145).

PANCREAS: ISLETS of LANGERHANS

Islets of Langerhans make up 1-2% of Pancreatic Tissue.

α cells — secrete ⟶ GLUCAGON
(25% of Islets)

BLOOD STREAM

promotes GLYCOGEN breakdown

increases Blood Sugar and utilization of GLUCOSE by tissues

Its effects are overshadowed by

x 150

β cells — secrete ⟶ INSULIN
(50-75% of Islets)   [In cells probably in combination with Zinc]

BLOOD STREAM

SECRETION of INSULIN appears to be stimulated primarily by a RISE in BLOOD SUGAR reaching Islets

Increases GLUCOSE UPTAKE by all "GLUCOSE-consuming tissues"- i.e. It plays important rôle in CARBOHYDRATE METABOLISM, probably regulating entrance of GLUCOSE to cells.

ANTERIOR PITUITARY
| ? Indirect action on
| Insulin production
GROWTH HORMONE ⎱
            ⎰ oppose action
SUPRARENAL   of Insulin
GLUCOCORTICOIDS ⎰

CHIEF ACTIONS on:-

KIDNEY TUBULES   stimulates reabsorption of GLUCOSE from FILTRATE

ATROPHY ⟶ DIABETES
of ISLETS   MELLITUS
            complex
            disorder
Absence of |
Insulin ⟶ HIGH BLOOD SUGAR ⟶
         SUGAR in URINE ⟶
         POLYURIA (since sugar lost
                   in solution in water)

LIVER reduces GLYCOGEN
       breakdown ⟶ GLUCOSE   } BLOOD
MUSCLES increases GLYCO-   SUGAR
         GEN formation   LOWERED
         from GLUCOSE

THIRST ⟶ POLYDIPSIA and POLYPHAGIA

Muscles are unable to use GLUCOSE efficiently ⟶ Great weakness.

Fat metabolized instead of Carbohydrate ⟋ Fat stores depleted ⟶ Wasting and loss of weight.
                                        ⟍ Ketone bodies in Blood and Urine.

If untreated ⟶ progressive ⟶ drowsiness ⟶ coma ⟶ death.

Excess Insulin (HYPERINSULINISM) ⟶ low blood sugar ⟶ irritability; sweating; hunger.
If untreated ⟶ reduction of metabolism of nervous tissues ⟶ giddiness ⟶ coma ⟶ death.

THYMUS

The Thymus is an irregularly-shaped organ lying behind the breast bone. It is relatively large in the child and reaches its maximum size at puberty. It closely resembles a Lymph Node.

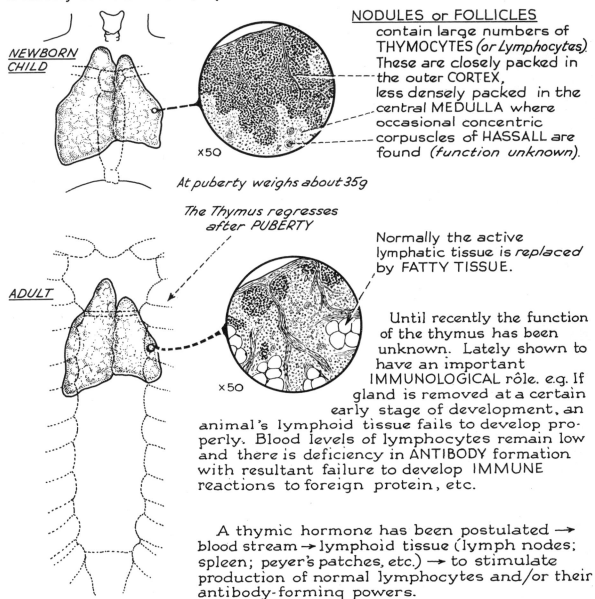

NEWBORN CHILD

x50

At puberty weighs about 35g

The Thymus regresses after PUBERTY

ADULT

x50

NODULES or FOLLICLES

contain large numbers of THYMOCYTES (or Lymphocytes). These are closely packed in the outer CORTEX, less densely packed in the central MEDULLA where occasional concentric corpuscles of HASSALL are found (function unknown).

Normally the active lymphatic tissue is *replaced* by FATTY TISSUE.

Until recently the function of the thymus has been unknown. Lately shown to have an important IMMUNOLOGICAL rôle. e.g. If gland is removed at a certain early stage of development, an animal's lymphoid tissue fails to develop properly. Blood levels of lymphocytes remain low and there is deficiency in ANTIBODY formation with resultant failure to develop IMMUNE reactions to foreign protein, etc.

A thymic hormone has been postulated → blood stream → lymphoid tissue (lymph nodes; spleen; peyer's patches, etc.) → to stimulate production of normal lymphocytes and/or their antibody-forming powers.

174

CHAPTER 8

REPRODUCTIVE SYSTEM

MALE REPRODUCTIVE SYSTEM

<u>PRIMARY SEX ORGANS</u>   *produce the MALE GERM CELLS –* <u>*SPERMATOZOA*</u>
TESTES (Two)   *and the MALE SEX HORMONE –*
                                        <u>*TESTOSTERONE*</u>

Testosterone is responsible for
development at Puberty of :–
<u>SECONDARY SEX ORGANS</u>
   EPIDIDYMIS (Two) ————⎤
   VAS DEFERENS (Two) ———⎱ *transfer Spermatozoa from the Testes.*
   SEMINAL VESICLES (Two) –⎱
   PROSTATE GLAND ————— *secrete fluid medium for transport of Spermatozoa.*
   PENIS ————————— *transfers Spermatozoa from male to female.*

   and
appearance of
<u>SECONDARY SEX CHARACTERISTICS</u>
Laryngeal changes → Deep voice.
Pubic, Axillary and Facial hair.
Characteristic Male shape of
body.

VAS DEFERENS ––
SEMINAL VESICLE ––
EPIDIDYMIS ––
TESTIS ––
SCROTUM ––
                    PENIS

In the male the process of spermatogenesis starts just after
puberty and is normally continuous until old age.

176

TESTIS

There are TWO TESTES.
These produce the Male GERM CELLS:–

SPERMATOGENESIS

The production of
SPERMATOZOA occurs in the
SEMINIFEROUS TUBULES.

SPERMATOGONIA divide *(by Mitosis)*
      to form
PRIMARY SPERMATOCYTES which
            divide *(by Meiosis)*
SECONDARY SPERMATOCYTES-divide
            (by Mitosis)
SPERMATIDS change gradually *(no
            further division)*

  free swimming
SPERMATOZOA *(approximately 0·1mm long)*

*[During development spermatozoa are
supported and nourished by
CELLS of SERTOLI.]*

Interstitial cells of LEYDIG produce the
<u>MALE SEX HORMONE</u> – *TESTOSTERONE* –

which passes directly into Blood Stream

to control development and activity of
SECONDARY SEX ORGANS. It is responsible
for appearance and maintenance of
SECONDARY SEX CHARACTERISTICS.

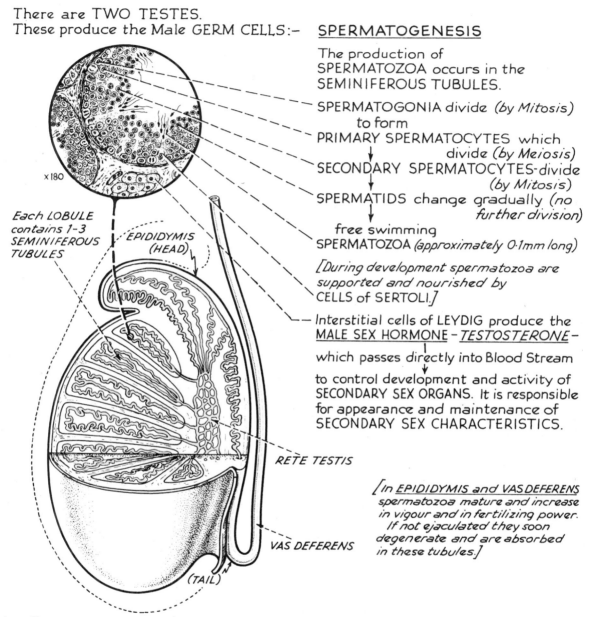

x180

*Each LOBULE
contains 1-3
SEMINIFEROUS
TUBULES*

*EPIDIDYMIS
(HEAD)*

RETE TESTIS

*[In <u>EPIDIDYMIS</u> and <u>VAS DEFERENS</u>
spermatozoa mature and increase
in vigour and in fertilizing power.
If not ejaculated they soon
degenerate and are absorbed
in these tubules.]*

VAS DEFERENS

(TAIL)

*Events occurring in the testes are under CONTROL of HORMONES chiefly
those of ANTERIOR PITUITARY.*

MALE SECONDARY SEX ORGANS

These are the organs adapted for TRANSFER of live SPERMATOZOA
from male to female.

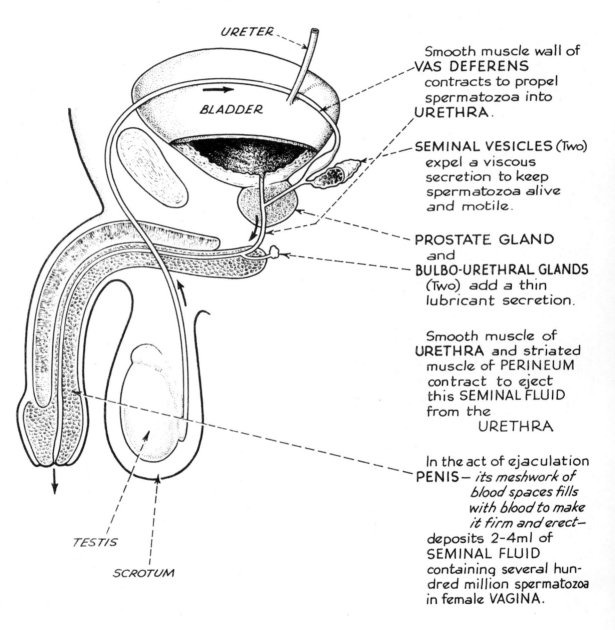

URETER

BLADDER

Smooth muscle wall of
VAS DEFERENS
contracts to propel
spermatozoa into
URETHRA.

SEMINAL VESICLES (Two)
expel a viscous
secretion to keep
spermatozoa alive
and motile.

PROSTATE GLAND
and
BULBO-URETHRAL GLANDS
(Two) add a thin
lubricant secretion.

Smooth muscle of
URETHRA and striated
muscle of PERINEUM
contract to eject
this SEMINAL FLUID
from the
URETHRA

In the act of ejaculation
PENIS — *its meshwork of*
blood spaces fills
with blood to make
it firm and erect —
deposits 2-4ml of
SEMINAL FLUID
containing several hun-
dred million spermatozoa
in female VAGINA.

TESTIS

SCROTUM

PUBERTY in MALE

Between the ages of 13 and 16 years Testicular tissue becomes responsive to stimulation by
ANTERIOR PITUITARY
GONADOTROPHIC HORMONES

A.P

Reciprocal Relationship

Reciprocal Relationship

FSH
(Follicle-Stimulating Hormone)
stimulates

SPERMATOGENESIS

LH *(Luteinizing Hormone)*
(or Interstitial Cell Stimulating Hormone)
stimulates secretion
and release of
TESTOSTERONE

BLOOD STREAM

TISSUES of BODY

causes appearance and
maintenance of signs
associated with PUBERTY:-
SECONDARY SEX
CHARACTERISTICS:
HAIR grows on FACE, AXILLAE,
ABDOMEN and PUBIS
LARYNX enlarges →
VOICE deepens

GROWTH: Skeletal
MUSCLES develop proportions
BONES grow   alter

SECONDARY SEX ORGANS
develop:

PENIS, SCROTUM,
PROSTATE enlarge

PSYCHOLOGICAL CHANGES
and SEXUAL INSTINCTS
associated with adulthood
begin to develop.

*Tubules and perhaps interstitial
cells apparently secrete small
amounts of Oestrogen. This
may regulate FSH output by
Anterior Pituitary*

*If Atrophy
of Testes
occurs* → *at time of normal puberty* ———→ Sex organs remain small.
→ Secondary sex characteristics fail
after puberty —→ Spermatogenesis stops ——→ Sterility.   to develop.
→ Testosterone production falls —→ Atrophy of secondary sex organs.

Injections of Testosterone in cases of delayed puberty → changes associated with puberty.

179

FEMALE REPRODUCTIVE SYSTEM

PRIMARY SEX ORGANS   *produce the FEMALE GERM CELLS — OVA*
    OVARIES (Two)   *and the FEMALE SEX HORMONES —*
                     OESTROGEN and PROGESTERONE

Oestrogen and Progesterone are responsible
for development at Puberty of:-
        SECONDARY SEX ORGANS

FALLOPIAN TUBES (Two) — for the transfer of the Ova from Ovaries.
VAGINA —————————— for the reception of the Male Germ Cells.
UTERUS ——————————— for the nutrition and development of the
                  fertilized Egg Cell → developing embryo.
MAMMARY GLANDS (Two)- for the nutrition of the New Individual after birth.

and
appearance and
maintenance of
SECONDARY SEX
CHARACTERISTICS

Development of
        Breasts

Axillary and Pubic
        hair

Typical Feminine
Proportions of body

FALLOPIAN
TUBES

MAMMARY
GLANDS

OVARY

UTERUS

VAGINA

In the female the cyclical production of ova starts just after puberty and
continues (unless interrupted by pregnancy or disease) until the menopause.

180

ADULT PELVIC SEX ORGANS
in ORDINARY FEMALE CYCLE

The UTERINE TUBE receives OVUM shed from OVARY at OVULATION

Peristaltic contractions of SMOOTH MUSCLE COATS aided by CILIARY MOVEMENT in CILIATED EPITHELIUM move OVUM towards UTERUS

×40

OVARY

LIGAMENT

OVARY

BROAD LIGAMENT

(ovum atrophies if fertilization does not occur)

UTERUS has an outer smooth muscle coat — the MYOMETRIUM and a specialized inner mucous membrane — the ENDOMETRIUM

This shows cyclical changes during each ordinary MENSTRUAL CYCLE

It DEGENERATES and is shed at MENSTRUATION

VAGINA is lined with modified skin — STRATIFIED SQUAMOUS EPITHELIUM — which shows cyclical changes (GLYCOGEN content is greatest towards end of Menstrual Cycle)

×40

Blood vessels become more coiled

REPAIR   GROWTH PHASE   SECRETORY PHASE

Shed with BLOOD and MUCUS in MENSTRUAL FLOW

28/1 DAY   3 4   14 DAY   28 th DAY

Rhythmical changes occur in UTERUS, UTERINE TUBES and VAGINA under action of OVARIAN Hormones.

OVARY in ORDINARY ADULT CYCLE

There are TWO OVARIES. These produce the Female GERM CELLS.
The production of OVA is a CYCLICAL PROCESS — OOGENESIS

① __FORMATION__

GERMINAL EPITHELIUM :
*in development groups of cells
migrate into* STROMA

② __GROWTH__
Central OOGONIUM
enlarges in PRIMORDIAL
FOLLICLE *to become*
PRIMARY OOCYTE
surrounded by
FOLLICULAR CELLS
*which multiply
to become*
CUMULUS OOPHORUS
*(attaching OVUM to
wall of follicle)
and* MEMBRANA
GRANULOSA
*which, together with
cells of Theca Interna,
secrete hormone*
__OESTROGEN__ *— partly
stored in* LIQUOR FOLLICULI
*in enlarging Cavity of
developing* GRAAFIAN FOLLICLE
*—partly absorbed into Blood
Vessels of* THECA INTERNA
↓
GENERAL CIRCULATION
↓
Controls changes in
SECONDARY SEX ORGANS
*in First Half of Menstrual
Cycle.*

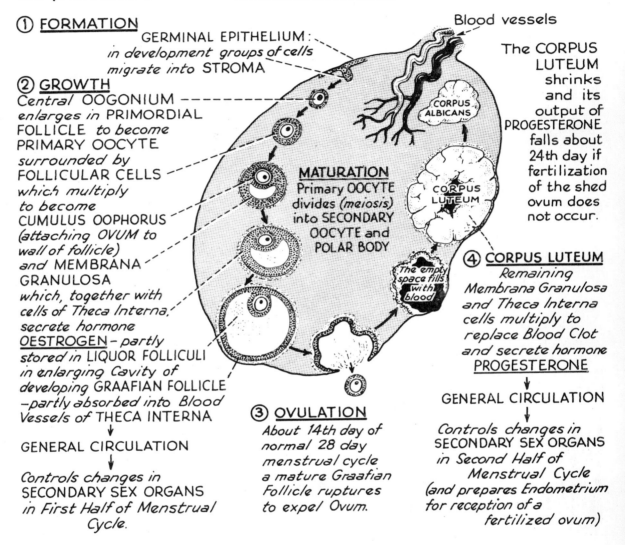

MATURATION
Primary OOCYTE
*divides (meiosis)
into* SECONDARY
OOCYTE *and*
POLAR BODY

*The empty
space fills
with
blood*

Blood vessels

The CORPUS
LUTEUM
*shrinks
and its
output of*
PROGESTERONE
*falls about
24th day if
fertilization
of the shed
ovum does
not occur.*

CORPUS ALBICANS

CORPUS LUTEUM

④ __CORPUS LUTEUM__
*Remaining
Membrana Granulosa
and Theca Interna
cells multiply to
replace Blood Clot
and secrete hormone*
__PROGESTERONE__
↓
GENERAL CIRCULATION
↓
Controls changes in
SECONDARY SEX ORGANS
*in Second Half of
Menstrual Cycle
(and prepares Endometrium
for reception of a
fertilized ovum)*

③ __OVULATION__
*About 14th day of
normal 28 day
menstrual cycle
a mature Graafian
Follicle ruptures
to expel Ovum.*

*For simplicity the development of only one Graafian Follicle is shown here.
Several grow in each cycle but in the Human subject usually only one Follicle
ruptures. The others atrophy: i.e. ONE MATURE OVUM is shed each month.
Events in the OVARY are under control of ANTERIOR PITUITARY HORMONES.*

OVARY in PREGNANCY

When pregnancy occurs the ordinary ovarian cycle is suspended.

The CORPUS LUTEUM continues to grow until it may come to occupy 30% to 50% of the total volume of the OVARY.

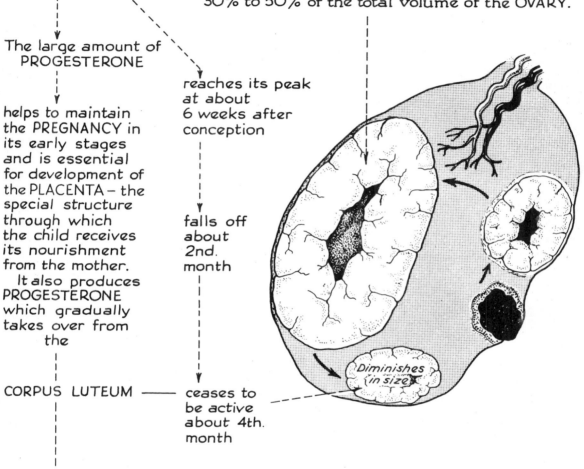

The large amount of PROGESTERONE

reaches its peak at about 6 weeks after conception

helps to maintain the PREGNANCY in its early stages and is essential for development of the PLACENTA – the special structure through which the child receives its nourishment from the mother.

It also produces PROGESTERONE which gradually takes over from the

falls off about 2nd. month

Diminishes in size

CORPUS LUTEUM —— ceases to be active about 4th. month

PLACENTAL PROGESTERONE —— takes over to maintain the PREGNANCY, and to help to prepare the mammary glands for Lactation.

PUBERTY in FEMALE

Between the ages of 10 and 14 years Ovarian tissue usually begins to respond to stimulation by <u>ANTERIOR PITUITARY GONADOTROPHIC HORMONES.</u>

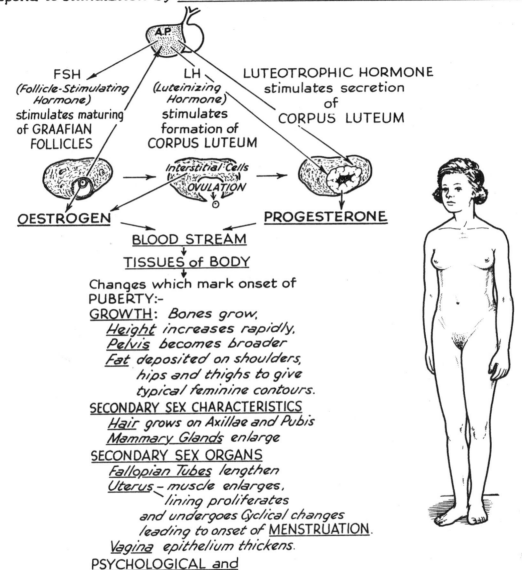

FSH
(Follicle-Stimulating Hormone)
stimulates maturing of GRAAFIAN FOLLICLES

LH
(Luteinizing Hormone)
stimulates formation of CORPUS LUTEUM

LUTEOTROPHIC HORMONE
stimulates secretion of CORPUS LUTEUM

Interstitial Cells

OVULATION

<u>OESTROGEN</u>

<u>PROGESTERONE</u>

<u>BLOOD STREAM</u>

<u>TISSUES of BODY</u>

Changes which mark onset of PUBERTY:-

<u>GROWTH</u>: *Bones grow,*
<u>Height</u> increases rapidly,
<u>Pelvis</u> becomes broader
<u>Fat</u> deposited on shoulders,
hips and thighs to give
typical feminine contours.

<u>SECONDARY SEX CHARACTERISTICS</u>
<u>Hair</u> grows on Axillae and Pubis
<u>Mammary Glands</u> enlarge

<u>SECONDARY SEX ORGANS</u>
<u>Fallopian Tubes</u> lengthen
<u>Uterus</u> — muscle enlarges,
lining proliferates
and undergoes Cyclical changes
leading to onset of <u>MENSTRUATION</u>.
<u>Vagina</u> epithelium thickens.

<u>PSYCHOLOGICAL</u> and
<u>PERSONALITY</u> *changes occur.*

i.e. PUBERTY changes GIRL CHILD into WOMAN able to bear children.

OVARIAN HORMONES

The Ovarian Cycle is repeated monthly from PUBERTY to the MENOPAUSE unless interrupted by PREGNANCY or DISEASE.

OVARIAN HORMONES are therefore directly responsible for regular cycle of events in SECONDARY SEX ORGANS.

UTERUS and UTERINE TUBES

The UTERUS (or Womb) is the organ which bears the developing child till birth.

In CHILDHOOD -

- it is a small undeveloped organ situated deep in the pelvis.

OUTER WALLS of
    Smooth muscle -
       the MYOMETRIUM

INNER LINING or
    Mucous membrane -
      the ENDOMETRIUM

At PUBERTY

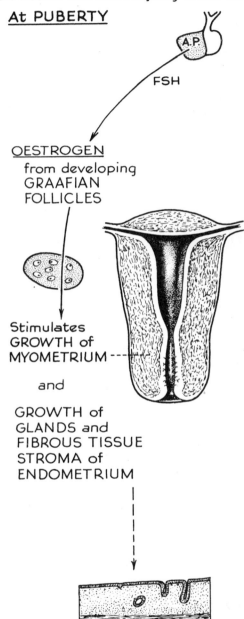

FSH

OESTROGEN
    from developing
    GRAAFIAN
    FOLLICLES

Stimulates
GROWTH of
MYOMETRIUM

and

GROWTH of
GLANDS and
FIBROUS TISSUE
STROMA of
ENDOMETRIUM

UTERUS and UTERINE TUBES

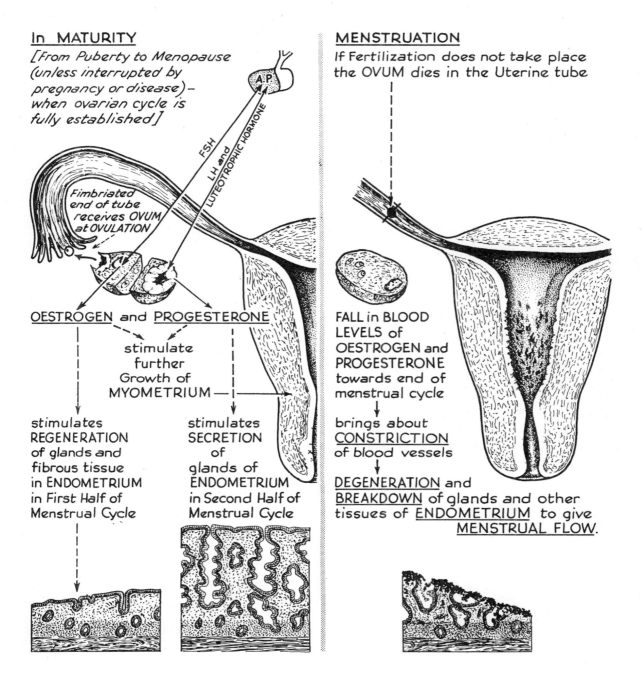

In MATURITY

[From Puberty to Menopause (unless interrupted by pregnancy or disease) – when ovarian cycle is fully established]

A.P.

FSH
LH and LUTEOTROPHIC HORMONE

Fimbriated end of tube receives OVUM at OVULATION

OESTROGEN and PROGESTERONE

stimulate further Growth of MYOMETRIUM—

stimulates REGENERATION of glands and fibrous tissue in ENDOMETRIUM in First Half of Menstrual Cycle

stimulates SECRETION of glands of ENDOMETRIUM in Second Half of Menstrual Cycle

MENSTRUATION

If Fertilization does not take place the OVUM dies in the Uterine tube

FALL in BLOOD LEVELS of OESTROGEN and PROGESTERONE towards end of menstrual cycle

brings about CONSTRICTION of blood vessels

DEGENERATION and BREAKDOWN of glands and other tissues of ENDOMETRIUM to give MENSTRUAL FLOW.

187

UTERINE TUBES
in cycle ending in PREGNANCY

The fimbriated end of the UTERINE TUBE receives the OVUM at OVULATION. PERISTALTIC contractions of the muscular tube aided by ciliary movements of its lining cells transfer the OVUM towards the UTERUS. The uterine tube also transmits SPERMATOZOA towards the OVA.

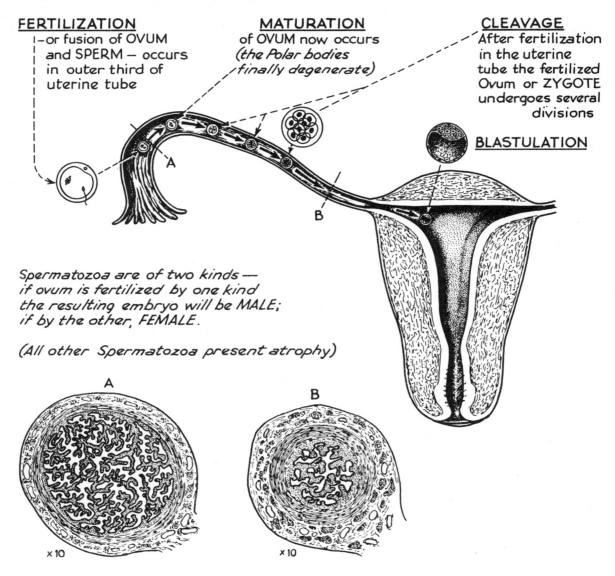

FERTILIZATION
—or fusion of OVUM and SPERM — occurs in outer third of uterine tube

MATURATION
of OVUM now occurs *(the Polar bodies finally degenerate)*

CLEAVAGE
After fertilization in the uterine tube the fertilized Ovum or ZYGOTE undergoes several divisions

BLASTULATION

A

B

Spermatozoa are of two kinds — if ovum is fertilized by one kind the resulting embryo will be MALE; if by the other, FEMALE.

(All other Spermatozoa present atrophy)

A

B

× 10

× 10

UTERUS

After FERTILIZATION

Developing embryo arrives
(about 4th to 7th day) in UTERUS
prepared for it by

Blood levels of
GONADOTROPHINS rise

OESTROGEN and PROGESTERONE

Embryo gets OXYGEN and
nutrients (*by diffusion*) from
this glandular secretion

Embryo sticks to lining of
WOMB

↓

Its surface TROPHOBLAST
cells fuse with, destroy
and finally penetrate the
ENDOMETRIUM

↓

Embryo now absorbs
TISSUE FLUIDS and
CELLULAR DEBRIS

↓

Outgrowths from
outer layer of
Blastocyst begin
to erode Endo-
metrium.

Ciliary currents
and PERISTALTIC
contractions in
UTERINE TUBE
    carry
BLASTOCYST into
UTERINE SECRETION

The ENDOMETRIUM is in LUTEAL PHASE
and continues to grow (*No menstrual*
degeneration occurs)
Glands are actively secreting mucus

Compact Stroma

Spongy layer with
widely dilated glands

Limiting layer—
glands tortuous

UTERUS

*Blood levels of
GONADOTROPHINS
continue to rise*

IMPLANTATION

Embryo invades
ENDOMETRIUM
*(which is now called
the DECIDUA)*

PLACENTATION

PROGESTERONE is necessary
for the development of the
PLACENTA – the special organ
through which the developing
child receives nourishment
from the mother

<u>OESTROGEN</u> and <u>PROGESTERONE</u>

*The HORMONES
of the PLACENTA
help to maintain
the
PREGNANCY*

*The Myometrial
Smooth muscle
cells increase
in number and
size*

*CHORIONIC VILLI
– Finger-like
projections from
the embryo have
invaded
mother's
endometrial
blood vessels*

*Blood vessels
develop in the
CHORIONIC VILLI
which are now
interlocked with
mother's tissues
and surrounded
by mother's
blood.
STRUCTURE formed
in this way is
the PLACENTA.*

PLACENTA

The PLACENTA in WOMAN is HAEMOCHORIAL – *i.e. chorionic villi dip directly into maternal blood.*
It increases in weight throughout Pregnancy.

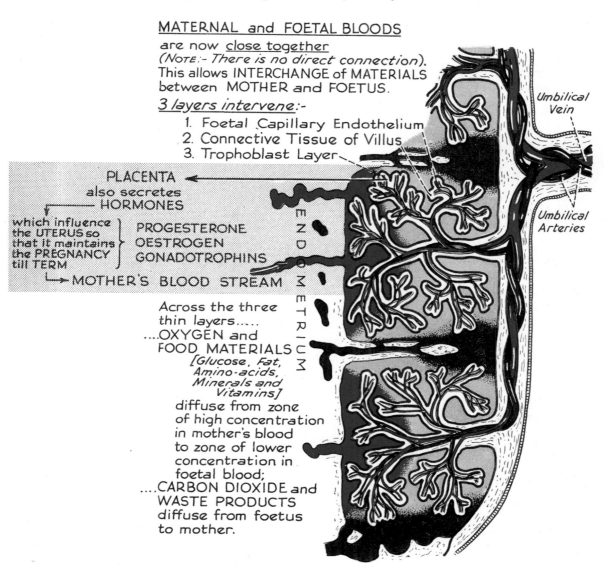

MATERNAL and FOETAL BLOODS
are now <u>close together</u>
(NOTE:- *There is no direct connection*).
This allows INTERCHANGE of MATERIALS between MOTHER and FOETUS.
3 layers intervene:-
1. Foetal Capillary Endothelium
2. Connective Tissue of Villus
3. Trophoblast Layer

PLACENTA
also secretes
HORMONES
which influence the UTERUS so that it maintains the PREGNANCY till TERM
PROGESTERONE
OESTROGEN
GONADOTROPHINS
MOTHER'S BLOOD STREAM

Across the three thin layers.....
...OXYGEN and FOOD MATERIALS
[Glucose, Fat, Amino-acids, Minerals and Vitamins]
diffuse from zone of high concentration in mother's blood to zone of lower concentration in foetal blood;
....CARBON DIOXIDE and WASTE PRODUCTS diffuse from foetus to mother.

ENDOMETRIUM

Umbilical Vein

Umbilical Arteries

Substances of small molecular weight usually pass in either direction by diffusion.
Larger molecules are probably transported by special carrier systems involving enzymes.

UTERUS

? *Role of Maternal Pituitary Gonadotrophins in maintaining pregnancy.*

Foetus grows larger and comes to fill UTERINE CAVITY

Corpus Luteum ceases to function after 3RD or 4TH month.
Suprarenal OESTROGEN and PROGESTERONE may play a part in maintenance of pregnancy.

Foetus is attached by UMBILICAL CORD to PLACENTA

Foetus is bathed by AMNIOTIC FLUID (within Amniotic and Chorionic membranes)

GROWTH of MYOMETRIUM —increase in number and size of smooth muscle cells and of the blood vessels

PLACENTAL OESTROGEN and PROGESTERONE GONADOTROPHINS in blood stream condition the maintenance of pregnancy

STRETCHING of MYOMETRIUM

FOETAL CIRCULATION

For the foetus the PLACENTA acts as the organ of transfer for Oxygen, Nutritives and Waste Products.   Only a small volume of blood passes through the foetal lungs.

BLOOD RETURNING TO HEART

...To *RIGHT ATRIUM*

Small amount from heart, head, neck and arms → *S.V.C.*

LARGE AMOUNT via UMBILICAL VEINS

through LIVER – short circuits to I.V.C. via
DUCTUS VENOSUS.

Some of this passes to Right Atrium
Small amount from abdominal cavity and legs.

...To *LEFT ATRIUM*

Small amount from 2 lungs

LARGE AMOUNT from INFERIOR VENA CAVA

through

FORAMEN OVALE.

(thus by-passing pulmonary circulation)

BLOOD LEAVING HEART

...*From RIGHT VENTRICLE*

Small amount to 2 lungs

LARGE AMOUNT to AORTA

through

DUCTUS ARTERIOSUS

(thus by-passing pulmonary circulation)
joins
OUTPUT from LEFT VENTRICLE

↓

Small amount to heart, head, neck and arms.
LARGE AMOUNT to PLACENTA

through
UMBILICAL ARTERIES

Small amount to abdominal cavity and legs.

FORAMEN OVALE
(Opening between inferior vena cava and left atrium)

DUCTUS VENOSUS

Umbilical Vein

LIVER

Umbilical Arteries

DUCTUS ARTERIOSUS

After G. S. Dawes

At birth the smooth muscle in the umbilical artery constricts to close it. Shortly after birth DUCTUS VENOSUS, DUCTUS ARTERIOSUS and FORAMEN OVALE close.

UTERUS

PARTURITION

About 40 weeks after conception the process of <u>CHILDBIRTH</u> usually begins ——— Factors initiating this are largely unknown.

1ST Stage usually lasts up to 14 hours with a first birth

<u>MYOMETRIUM</u>

Uterine muscle is now very greatly distended – *[this distension is probably not an important factor in initiating Parturition].*

Rhythmic contractions begin to increase in strength and frequency

Press on Amniotic Fluid

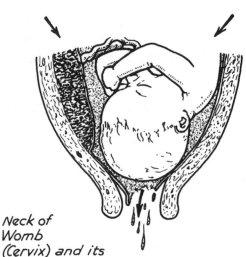

Neck of Womb (Cervix) and its opening (Os) dilate

Membranes rupture and <u>AMNIOTIC FLUID</u> escapes

After 32ND week of Pregnancy the uterus becomes sensitive to OXYTOCIN —— its importance in Parturition is unknown.

2ND Stage LABOUR usually lasts up to 2 hours with a first birth

Uterine contractions increase in strength and frequency and (aided by voluntary contractions of Abdominal muscles)

CHILD is slowly forced through PELVIS

BIRTH of BABY

UTERUS

After PARTURITION

3RD Stage Labour
5-15 minutes
after birth of child

HYPOTHALAMUS

<u>OXYTOCIN</u>
released to blood stream
stimulates
CONTRACTIONS of UTERINE
MUSCLE

DETACH and DELIVER PLACENTA
and the membranes as the
<u>AFTERBIRTH</u>.

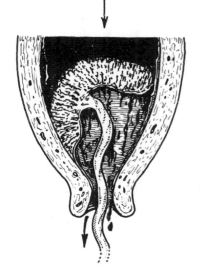

In PUERPERIUM

Immediately
following childbirth.

*-- Afferents
from Breasts
in Suckling*

P.P.

FALL in
BLOOD LEVELS
of
OESTROGEN,
PROGESTERONE
and
GONADOTROPHINS
(after loss of Placenta)

OXYTOCIN
released to
blood stream

<u>MYOMETRIUM</u>

Uterine muscle contracts down to
close off blood vessels torn and
bleeding after separation of
Placenta.

A large part of Endometrium —
Decidua — is shed with the
Placenta.
Only the limiting layer is left.

UTERUS

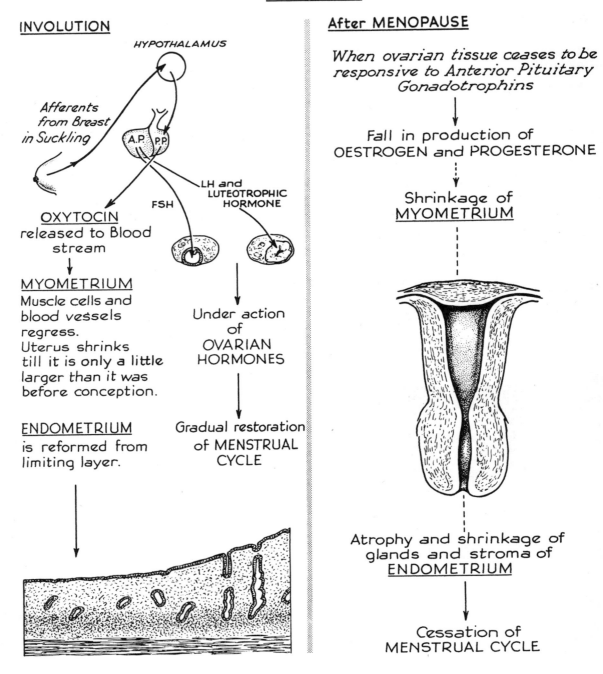

INVOLUTION

HYPOTHALAMUS

Afferents
from Breast
in Suckling

A.P. P.P.

LH and
LUTEOTROPHIC
HORMONE

FSH

OXYTOCIN
released to Blood
stream

MYOMETRIUM
Muscle cells and
blood vessels
regress.
Uterus shrinks
till it is only a little
larger than it was
before conception.

ENDOMETRIUM
is reformed from
limiting layer.

Under action
of
OVARIAN
HORMONES

Gradual restoration
of MENSTRUAL
CYCLE

After MENOPAUSE

When ovarian tissue ceases to be
responsive to Anterior Pituitary
Gonadotrophins

Fall in production of
OESTROGEN and PROGESTERONE

Shrinkage of
MYOMETRIUM

Atrophy and shrinkage of
glands and stroma of
ENDOMETRIUM

Cessation of
MENSTRUAL CYCLE

MAMMARY GLANDS

There are 2 mammary glands.

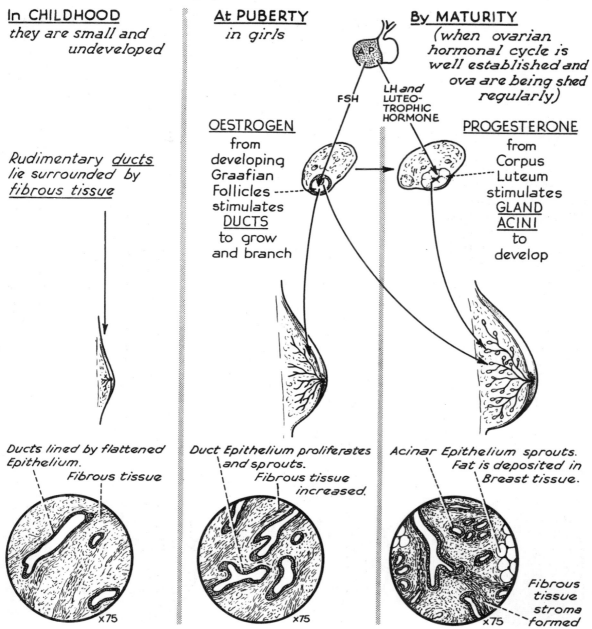

In CHILDHOOD
they are small and undeveloped

At PUBERTY
in girls

By MATURITY
(when ovarian hormonal cycle is well established and ova are being shed regularly)

FSH

LH and LUTEO-TROPHIC HORMONE

Rudimentary ducts lie surrounded by fibrous tissue

OESTROGEN
from developing Graafian Follicles ---- stimulates DUCTS to grow and branch

PROGESTERONE
from Corpus Luteum stimulates GLAND ACINI to develop

Ducts lined by flattened Epithelium.
Fibrous tissue

Duct Epithelium proliferates and sprouts.
Fibrous tissue increased.

Acinar Epithelium sprouts.
Fat is deposited in Breast tissue.

Fibrous tissue stroma formed

x75

x75

x75

MAMMARY GLANDS

In PREGNANCY

PROGESTERONE
from
Corpus Luteum
and
OESTROGEN and
PROGESTERONE
from Placenta
stimulate
further GROWTH
of DUCTS and
ACINI.

AREOLA
and
NIPPLE
becomes
darkly
PIGMENTED

Acinar Epithelium becomes cuboidal

×75

After CHILDBIRTH

Fall in Blood Level
of OESTROGEN
and PROGESTERONE
(after loss of Placenta)
causes release of
LUTEOTROPHIC HORMONE
(PROLACTIN)
from
ANTERIOR
PITUITARY

stimulates
prepared
GLAND ACINI
to secrete
MILK

Constituents
of MILK are
derived
from blood
flowing
through
gland

Apocrine Acini — luminal
margin
breaks
down
to
release
secretion

×75

Emotional factors influence LACTATION

HYPO-THALAMUS

PROLACTIN (LUTEOTROPHIN)

OXYTOCIN

Milk production
starts 3-4 days
after childbirth
and is maintained
(often for several
months) by
PROLACTIN.
It is influenced
by Post. Pituitary
OXYTOCIN.
AFFERENT nerve
impulses, set up
by Child SUCKLING,
give rise to reflex
EFFERENT nerve
impulses to Post.
Pituitary for the
release of OXYTOCIN
- carried by blood
stream to GLAND
ACINI
to
maintain
secretion
of MILK.

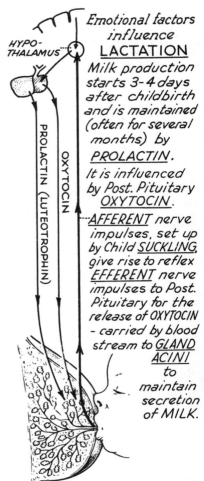

Globules of Fat in Acinar cells
collect and
coalesce →
cells burst
→ milk
discharged
to lumen
of duct.

×75

198

MAMMARY GLANDS

POST LACTATION

FSH

LH and
LUTEOTROPHIC
HORMONE

GRAAFIAN
FOLLICLES

CORPUS
LUTEUM

Re-establishment
of Ovarian Cycle:
Oestrogen and
Progesterone

Regression of
Secreting Acini
and Duct Tissue
till only little
more present
than before
pregnancy

x 75

AT MENOPAUSE

FSH

LH and
LUTEOTROPHIC
HORMONE

*Ovarian tissue becomes
less sensitive*

Withdrawal of
Oestrogen and
Progesterone

Shrinkage of
Breast Tissue
Acini
Ducts

x 75

MENOPAUSE

Between the ages of 42 and 50 years OVARIAN tissue gradually ceases to respond to stimulation by ANTERIOR PITUITARY GONADO-TROPHIC HORMONES.

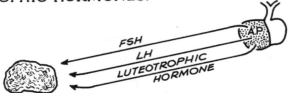

OVARIAN CYCLE becomes irregular and finally ceases ⟶ Ovary becomes small and fibrosed and no longer produces ripe Ova.

OESTROGEN and PROGESTERONE levels in Blood stream fall.

TISSUES of the body ——— begin to show changes which mark the end of REPRODUCTIVE LIFE.

Sometimes final redistribution of fat → less typically feminine distribution.

Regression of *Secondary Sex Characteristics*.

Breasts shrink.

Ducts⎫ Acini⎭ Atrophy

Hair becomes sparse in axillae and on pubis.

Secondary Sex Organs atrophy.

Fallopian tubes shrink.

Uterine Cycle and Menstruation cease.

(Muscle and lining shrink).

Vaginal epithelium becomes thin.

External Genitalia shrink.

Psychological and Personality changes.

Decline in Sexual powers.

Emotional disturbances may occur – often accompanied by Vasomotor phenomena such as "Hot Flushes" (vasodilatation), excessive sweating and giddiness.

Uterus

Vagina -

After the Menopause a woman is usually unable to bear children.

RELATIONSHIP between ANTERIOR PITUITARY OVARIAN and ENDOMETRIAL CYCLES

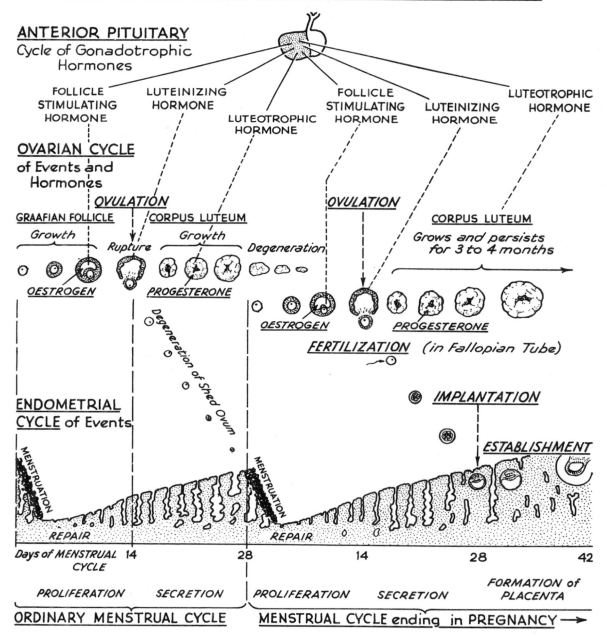

ANTERIOR PITUITARY
Cycle of Gonadotrophic
Hormones

FOLLICLE
STIMULATING
HORMONE

LUTEINIZING
HORMONE

LUTEOTROPHIC
HORMONE

FOLLICLE
STIMULATING
HORMONE

LUTEINIZING
HORMONE

LUTEOTROPHIC
HORMONE

OVARIAN CYCLE
of Events and
Hormones

OVULATION

OVULATION

GRAAFIAN FOLLICLE

Growth

Rupture

CORPUS LUTEUM

Growth

Degeneration

CORPUS LUTEUM

*Grows and persists
for 3 to 4 months*

OESTROGEN

PROGESTERONE

Degeneration of Shed Ovum

OESTROGEN

PROGESTERONE

FERTILIZATION (in Fallopian Tube)

IMPLANTATION

**ENDOMETRIAL
CYCLE of Events**

ESTABLISHMENT

MENSTRUATION

MENSTRUATION

REPAIR

REPAIR

Days of MENSTRUAL 14
CYCLE

28

14

28

42

PROLIFERATION   SECRETION

PROLIFERATION   SECRETION

FORMATION of
PLACENTA

ORDINARY MENSTRUAL CYCLE

MENSTRUAL CYCLE ending in PREGNANCY →

CHAPTER 9

"MASTER" TISSUES:

CENTRAL NERVOUS SYSTEM
LOCOMOTOR SYSTEM

NERVOUS SYSTEM

The Nervous System is concerned with the INTEGRATION and CONTROL of all bodily functions.

It has specialized in IRRITABILITY— *the ability to receive and respond to messages from the external and internal environments*

and also in CONDUCTION — *the ability to transmit messages to and from CO-ORDINATING CENTRES.*

The NERVOUS SYSTEM
consists of
a

↓

CENTRAL PART —
 The *BRAIN* and
SPINAL CORD

↓

linked by an
outlying
or

↓

PERIPHERAL PART — Nerve fibres

↓

TISSUES and ORGANS
of the body

SENSORY Nerve fibres carry messages from Tissues and Organs to the Brain or Spinal Cord.

MOTOR Nerve fibres carry messages to Tissues and Organs from the Brain or Spinal Cord.

DEVELOPMENT of the NERVOUS SYSTEM

The Nervous System develops in the embryo from a simple tube of ECTODERM:- The PRIMITIVE NEURAL TUBE.

The CELLS lining it become the NERVOUS tissue of the BRAIN and SPINAL CORD. The CANAL becomes distended to form the VENTRICLES of the BRAIN and CENTRAL CANAL of the SPINAL CORD:-

FORE MID HIND

Prosencephalon   Mesencephalon   Rhombencephalon

Each of these swellings with the cells lining them become more complicated:-

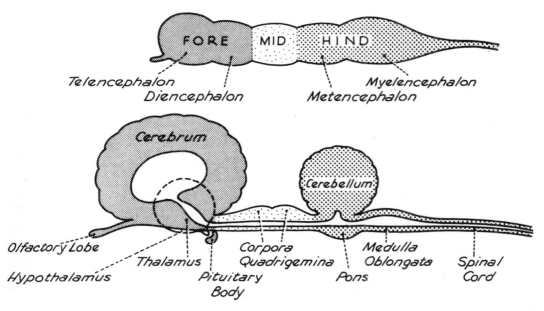

FORE MID HIND

Telencephalon
Diencephalon
Metencephalon
Myelencephalon

Cerebrum

Cerebellum

Olfactory Lobe
Hypothalamus
Thalamus
Pituitary Body
Corpora Quadrigemina
Pons
Medulla Oblongata
Spinal Cord

There are now many layers of cells forming the Brain and Spinal Cord.   The Ventricles of the Brain and the Central Canal of the Spinal Cord are filled with CEREBRO-SPINAL FLUID.

CEREBRUM

The largest part of the human brain is the CEREBRUM — made up of 2 CEREBRAL HEMISPHERES. Each of these is divided into LOBES.

INITIATING CENTRES for OUTGOING messages

RECEIVING CENTRES for INCOMING information

CENTRAL FISSURE

LONGITUDINAL FISSURE

PARIETAL LOBE

MOTOR AREA

PREMOTOR ASSOCIATION AREA

SENSORY AREA (Body)

SENSORY ASSOCIATION AREA

OCCIPITAL LOBE

Vision

Vision Association Area

FRONTAL LOBE

LATERAL FISSURE

Hearing

TEMPORAL LOBE

Motor Speech Area (one- on left side only)

Hearing Association Area

MOTOR CORTEX

GREY MATTER

BETZ CELL

WHITE MATTER

Large uncharted areas of the Cerebral Hemispheres are probably concerned with MENTAL PROCESSES such as *Intelligence, Memory, Judgement, Imagination, Creative* and *Conscious Thought.*

The surface of the brain shows many folds or CONVOLUTIONS. This has the effect of increasing the amount of GREY MATTER present. The GREY MATTER forms the outer layer or CORTEX. It contains the cell bodies of the NEURONES arranged in many interconnecting layers to form a 3-dimensional network.

About 90% of all Nerve Cells are in the Cerebral Cortex.

HORIZONTAL SECTION through BRAIN

This view shows surface GREY MATTER containing Nerve Cells and inner WHITE MATTER made up of Nerve Fibres.

Deep in the substance of the Cerebral Hemispheres there are additional masses of GREY MATTER :-

The BASAL GANGLIA   and THALAMUS

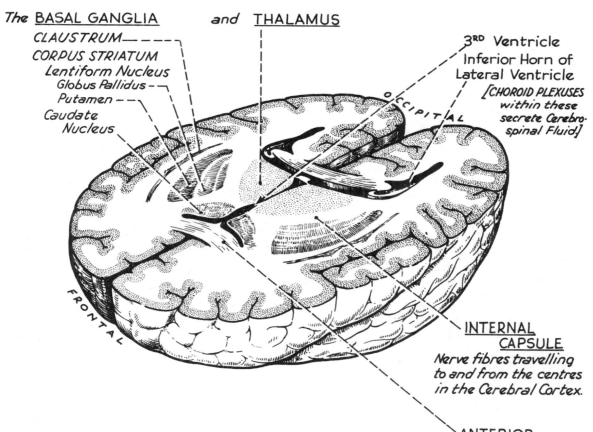

The Basal Ganglia
   are concerned with
   modifying and
   co-ordinating
   VOLUNTARY MUSCLE
   MOVEMENT.

The Thalamus
   is an important
   relay centre for
   SENSORY fibres
   on their way to
   CEREBRAL CORTEX.
   'Crude' Sensation
   and PAIN may be
   appreciated here.

CRANIAL NERVES

Twelve pairs of nerves arise directly from the undersurface of the Brain to supply Head and Neck and most of the viscera.

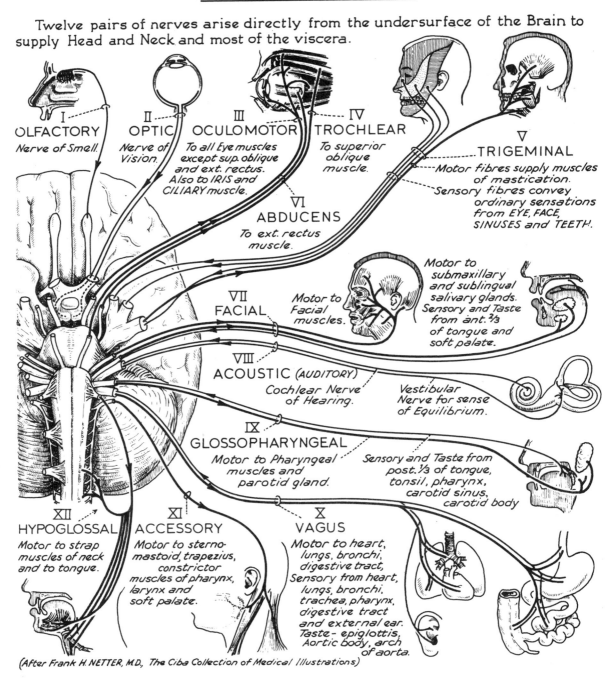

I
OLFACTORY
Nerve of Smell.

II
OPTIC
Nerve of Vision.

III
OCULOMOTOR
To all Eye muscles except sup. oblique and ext. rectus. Also to IRIS and CILIARY muscle.

IV
TROCHLEAR
To superior oblique muscle.

V
TRIGEMINAL
Motor fibres supply muscles of mastication. Sensory fibres convey ordinary sensations from EYE, FACE, SINUSES and TEETH.

VI
ABDUCENS
To ext. rectus muscle.

VII
FACIAL
Motor to Facial muscles.
Motor to submaxillary and sublingual salivary glands. Sensory and Taste from ant. ⅔ of tongue and soft palate.

VIII
ACOUSTIC (AUDITORY)
Cochlear Nerve of Hearing.
Vestibular Nerve for sense of Equilibrium.

IX
GLOSSOPHARYNGEAL
Motor to Pharyngeal muscles and parotid gland.
Sensory and Taste from post. ⅓ of tongue, tonsil, pharynx, carotid sinus, carotid body

XII
HYPOGLOSSAL
Motor to strap muscles of neck and to tongue.

XI
ACCESSORY
Motor to sterno-mastoid, trapezius, constrictor muscles of pharynx, larynx and soft palate.

X
VAGUS
Motor to heart, lungs, bronchi, digestive tract, Sensory from heart, lungs, bronchi, trachea, pharynx, digestive tract and external ear. Taste - epiglottis, Aortic body, arch of aorta.

(After Frank H. NETTER, M.D., The Ciba Collection of Medical Illustrations)

210

SPINAL CORD

The SPINAL CORD lies within the Vertebral Canal. It is continuous above with the Medulla Oblongata.

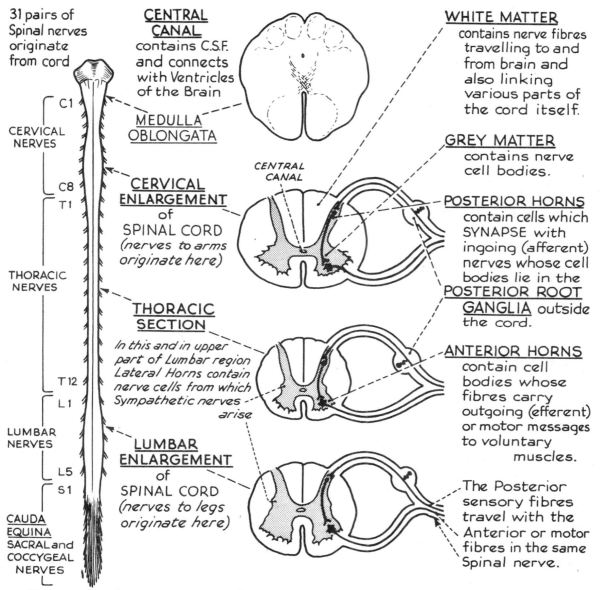

31 pairs of Spinal nerves originate from cord

CERVICAL NERVES

C1

C8

T1

THORACIC NERVES

T12

L1

LUMBAR NERVES

L5

S1

CAUDA EQUINA
SACRAL and COCCYGEAL NERVES

CENTRAL CANAL contains C.S.F. and connects with Ventricles of the Brain

MEDULLA OBLONGATA

CERVICAL ENLARGEMENT of SPINAL CORD *(nerves to arms originate here)*

THORACIC SECTION

In this and in upper part of Lumbar region Lateral Horns contain nerve cells from which Sympathetic nerves arise

LUMBAR ENLARGEMENT of SPINAL CORD *(nerves to legs originate here)*

CENTRAL CANAL

WHITE MATTER contains nerve fibres travelling to and from brain and also linking various parts of the cord itself.

GREY MATTER contains nerve cell bodies.

POSTERIOR HORNS contain cells which SYNAPSE with ingoing (afferent) nerves whose cell bodies lie in the **POSTERIOR ROOT GANGLIA** outside the cord.

ANTERIOR HORNS contain cell bodies whose fibres carry outgoing (efferent) or motor messages to voluntary muscles.

The Posterior sensory fibres travel with the Anterior or motor fibres in the same Spinal nerve.

The spinal nerves travel to all parts of the trunk and limbs.

211

SYNAPSE

The structural unit of the Nervous System is the NEURONE.
Neurones are linked together in the Nervous System.....

.....Nerve Process to Nerve Cell body at a SYNAPSE

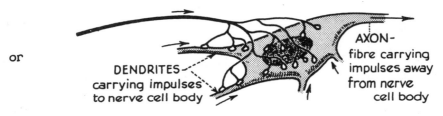

or

DENDRITES
carrying impulses
to nerve cell body

AXON -
fibre carrying
impulses away
from nerve
cell body

AXON ends in small
swellings – END FEET-
which merely touch
the DENDRITES or
BODY of another
NERVE CELL.
i.e. There is no direct
protoplasmic union
between neurones
at the SYNAPSE.

.....Nerve Process to Nerve Process at a SYNAPSE

One neurone usually connects with a great many others often widely
scattered in different parts of the Brain and Spinal Cord. In this way intricate
chains of nerve cells forming complex pathways for _incoming_ and _outgoing_
information can be built up within the Central Nervous System.

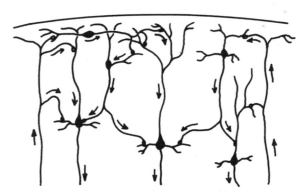

When the NERVE IMPULSE – like a very small electrical current – reaches a
SYNAPSE it causes the release at the NERVE ENDINGS of a CHEMICAL SUBSTANCE
which bridges the gap and forms the stimulus to conduct the Impulse to
the next Neurone.
A Synapse permits transmission of the impulse in one direction only.

212

NERVE IMPULSE

Neurones specialize in IRRITABILITY and in CONDUCTION of IMPULSES.

In <u>*RESTING (INACTIVE) NERVE FIBRE*</u>*:*

Surface membrane is permeable to   impermeable to (Na⁺)

The inside is electrically negative to the outside

The outside is electrically positive

+ + + + + + + + +

This is the so-called 'POLARIZED' or resting STATE. It is the result of chemical processes in the cells and is a potential source of energy.

When <u>*NERVE FIBRE is STIMULATED*</u> *(e.g. electrically or chemically)* ELECTRO-CHEMICAL CHANGES occur at point of stimulation. The resting potential is abolished.

The membrane is said to be 'DEPOLARIZED'.

The membrane permeability is altered

The electrical state is reversed. The inside becomes electrically positive and the outside becomes electrically negative.

Current flows ahead from the stimulated or active part to the inactive part – carried largely by the Potassium and Sodium ions.

Current passing through the membrane of the inactive part in turn depolarizes it. This then becomes the new active area to excite the next segment.

Direction of Impulse

In <u>*MYELINATED NERVE FIBRE*</u>*:*

The active region still 'triggers' the resting part ahead of it by causing an outward flow of current but this occurs only at the Nodes of Ranvier. i.e. The Nerve Impulse is propagated from Node to Node.

This increases speed of CONDUCTION along these large fibres.

After passage of nerve impulse 'REPOLARIZATION' and IONIC BALANCE are quickly restored.

A nerve impulse is the same whether it is in a sensory nerve taking messages to the brain or in a motor nerve taking impulses from the brain to the muscles, etc.

REFLEX ACTION

The NEURONE is the ANATOMICAL or STRUCTURAL UNIT of the Nervous System: the Nervous REFLEX is the PHYSIOLOGICAL or FUNCTIONAL UNIT.

A *Nervous Reflex* is an INVOLUNTARY ACTION caused by the STIMULATION of an AFFERENT *(sensory)* nerve ending or RECEPTOR.

The structural basis of Reflex Action is the REFLEX ARC. In its simplest form this consists of :-

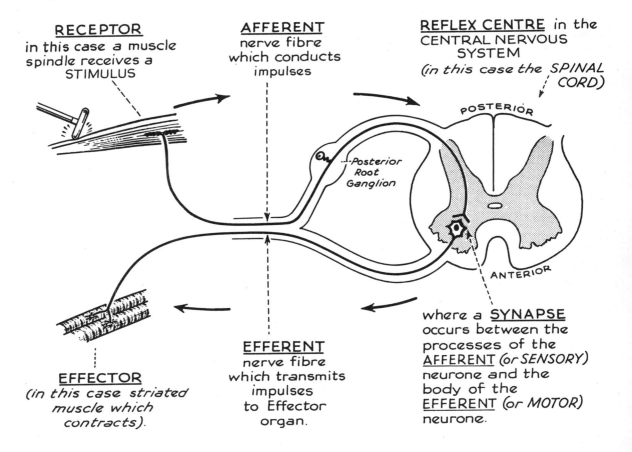

RECEPTOR
in this case a muscle spindle receives a STIMULUS

AFFERENT
nerve fibre which conducts impulses

REFLEX CENTRE in the CENTRAL NERVOUS SYSTEM
(in this case the SPINAL CORD)

POSTERIOR

Posterior Root Ganglion

ANTERIOR

EFFECTOR
(in this case striated muscle which contracts).

EFFERENT
nerve fibre which transmits impulses to Effector organ.

where a **SYNAPSE** occurs between the processes of the **AFFERENT** *(or SENSORY)* neurone and the body of the **EFFERENT** *(or MOTOR)* neurone.

Reflexes form the basis of all Central Nervous System (CNS) activity. They occur at all levels of the Brain and Spinal Cord. Important bodily functions such as movements of Respiration, Digestion, etc, are all controlled through Reflexes. We are made aware of some reflex acts: others occur without our knowledge.

STRETCH REFLEXES

In man a very few REFLEX ARCS involve 2 NEURONES only.
Two examples elicited by doctors when testing the Nervous System
are:—

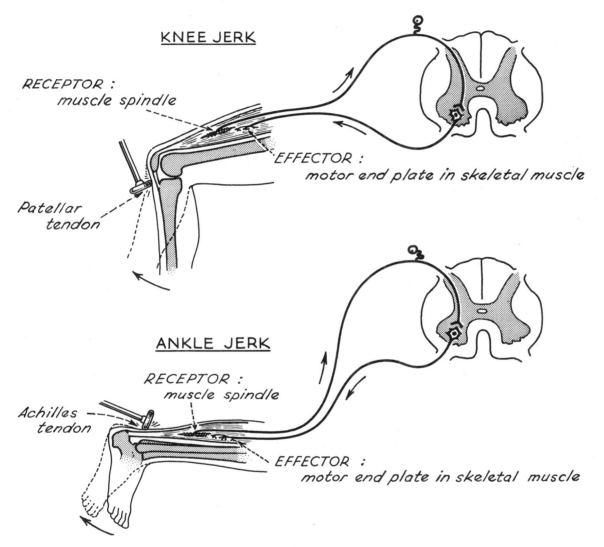

KNEE JERK

RECEPTOR :
muscle spindle

EFFECTOR :
motor end plate in skeletal muscle

Patellar
tendon

ANKLE JERK

RECEPTOR :
muscle spindle

Achilles
tendon

EFFECTOR :
motor end plate in skeletal muscle

When the tendon is sharply tapped the muscle is stretched. (N.B. The
stimulus is by _stretch_ of the _muscle spindle_.) Nerve messages pass
into the spinal cord — and out to the muscle which then contracts.

SPINAL REFLEXES

In most REFLEX ARCS in man AFFERENT and EFFERENT
NEURONES are linked by at least 1 CONNECTOR NEURONE.

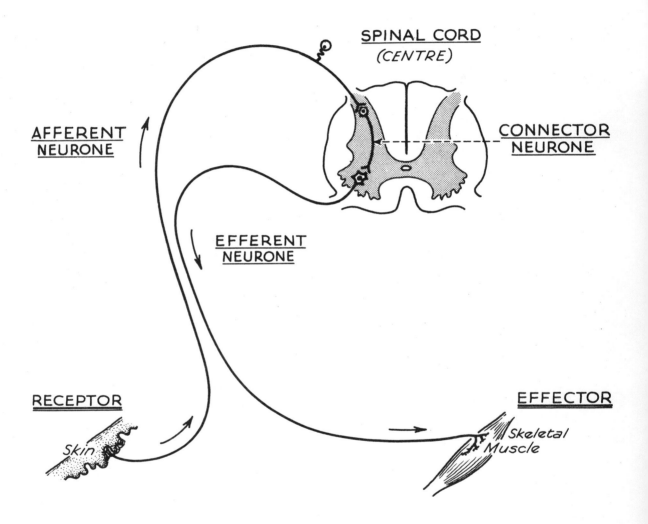

A chain of many connector neurones is frequently found.

"EDIFICE" of the C.N.S. *(After R.C. GARRY)*

In the majority of Reflex arcs in man a chain of many connector neurones is found. There may be link-ups with various levels of the Brain and Spinal Cord.

This diagram gives a highly simplified concept of the type of LINK-UP which can occur between different levels of the Central Nervous System.

At the LEVEL of the CEREBRAL CORTEX

Highest Sensory Centres

Third Sensory Neurone — SENSORY CORTEX   MOTOR CORTEX

PYRAMIDAL TRACT or Upper Motor Neurone

Highest Motor Centres

At the LEVEL of the BRAIN STEM

THALAMUS

NEURONES of EXTRAPYRAMIDAL SYSTEM e.g.

CORPUS STRIATUM
RED NUCLEUS
VESTIBULAR NUCLEUS

Second Sensory Neurone

CEREBELLUM

Lower Sensory Centres

SPINO-CEREBELLAR TRACT

Lower Motor Centres

MEDULLA

PART OF EXTRAPYRAMIDAL TRACT

At the LEVEL of the SPINAL CORD

First Sensory (or ingoing) Neurone

POSTERIOR HORN   ANTERIOR HORN

CONNECTOR NEURONE

MOTOR POOL

Lower Motor (or outgoing) Neurone

RECEPTOR   CENTRAL NERVOUS SYSTEM (C.N.S.)   **EFFECTOR**

Every RECEPTOR neurone is thus potentially linked in the C.N.S. with a large number of EFFECTOR organs all over the body and every EFFECTOR neurone is similarly in communication with RECEPTORS all over the body.

Centres in the BRAIN and BRAIN STEM can thus modify REFLEX ACTS which occur through the SPINAL CORD. These centres can send "suppressing" or "facilitating" impulses along their pathways to the cells in the SPINAL CORD.

REFLEX ACTION

Most REFLEX ACTIONS in man involve a great many REFLEX ARCS.

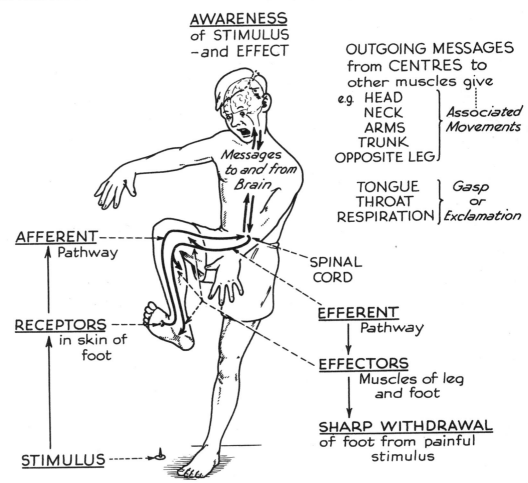

AWARENESS
of STIMULUS
–and EFFECT

OUTGOING MESSAGES
from CENTRES to
other muscles give

e.g. HEAD
NECK
ARMS
TRUNK
OPPOSITE LEG
} *Associated Movements*

TONGUE
THROAT
RESPIRATION
} *Gasp or Exclamation*

Messages to and from Brain

AFFERENT
Pathway

RECEPTORS
in skin of
foot

STIMULUS

SPINAL
CORD

EFFERENT
Pathway

EFFECTORS
Muscles of leg
and foot

SHARP WITHDRAWAL
of foot from painful
stimulus

The localized stimulation of a very few RECEPTORS – sending messages along their AFFERENT NEURONES to SPINAL CORD and to BRAIN – has led to a very large number of outgoing impulses in many EFFECTOR NEURONES to a large number of MUSCLES to give a very widespread and generalized REFLEX RESPONSE.

This is possible because each receptor neurone is potentially connected within the Central Nervous System with all effector neurones.

218

ARRANGEMENT of NEURONES

Some of the ways in which neurones can be linked are indicated here:-

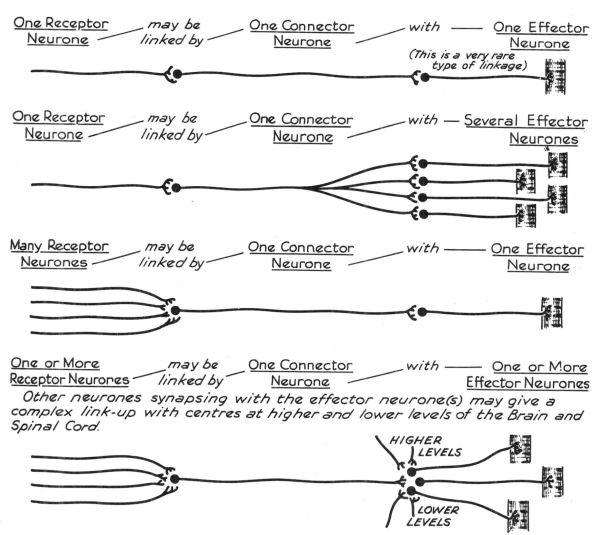

One Receptor Neurone — may be linked by — One Connector Neurone — with — One Effector Neurone
(This is a very rare type of linkage)

One Receptor Neurone — may be linked by — One Connector Neurone — with — Several Effector Neurones

Many Receptor Neurones — may be linked by — One Connector Neurone — with — One Effector Neurone

One or More Receptor Neurones — may be linked by — One Connector Neurone — with — One or More Effector Neurones

Other neurones synapsing with the effector neurone(s) may give a complex link-up with centres at higher and lower levels of the Brain and Spinal Cord.

HIGHER LEVELS

LOWER LEVELS

Through such 'functional' link-ups, neurones in different parts of the Central Nervous System, when active, can influence each other. This makes it possible for 'CONDITIONED' REFLEXES to become established (for simple example see page 55).
Such reflexes probably form the basis of all training so that it becomes difficult to say where REFLEX (or INVOLUNTARY) behaviour ends and purely VOLUNTARY behaviour begins.

219

SENSE ORGANS

Man's awareness of the world is limited to those forms of energy, physical or chemical, to which he has Receptors designed to respond. (Many "events" in the universe go unnoticed by man because he has no Sense Organ which can respond to them.)

Each Sense Organ is designed to respond to one type of stimulation. EXTEROCEPTORS are stimulated by events in the EXTERNAL ENVIRONMENT.

MECHANICAL VIBRATIONS caused by object vibrating (The disturbance of molecules in atmosphere travels as waves – measured in double vibrations per second)

20,000 d.v./sec.

several 100

20

0 d.v./sec.

SOUND WAVES — stimulate receptors in EAR

PRESSURE WAVES — stimulate endings in SKIN

Various physical or mechanical forces form CONTACT stimuli which lead to depolarization and generate nerve impulses from special sense organs in skin.

CHEMICAL CHANGES in ENVIRONMENT
—Chemical substances in solution
—Gaseous substances in solution

EYES
NOSE
TONGUE

stimulate HEAT receptors in skin

widespread stimulation of sense organs

ELECTROMAGNETIC WAVES (wavelengths measured in Ångström units: 1Å = one ten-millionth of a millimetre)

·000001
·003
1·4
250
3900
7800

VISIBLE RAYS

COSMIC
GAMMA
X-RAYS
ULTRA-VIOLET
INFRA-RED
RADIO
DOMESTIC ELECTRICAL POWER WAVES

All have chemical action on body's tissues – e.g. radiation of atomic energy 'kills' tissues.

'Sunburn'

2,000,000 Å (·2mm)

1,000's of METRES (KILOMETRES)
6,000,000 METRES

Exteroceptors may convey information to Consciousness with Awareness or Sensation and lead to suitable Responses planned in Cerebral Cortex or they may serve as Afferent pathways for Reflex (or Involuntary) Action with or without rising to consciousness.

PROPRIOCEPTORS are stimulated by changes in LOCOMOTOR SYSTEM of body

LABYRINTH ----- movements and position of head
MUSCLES -------- stretch
TENDONS ------- tension and stretch
JOINTS ---------- stretch and pressure

SENSE of EQUILIBRIUM or BALANCE and AWARENESS of POSITION and MOVEMENT of BODY in space.

INTEROCEPTORS in VISCERA are stimulated by changes in INTERNAL ENVIRONMENT (e.g. by distension in hollow organs).

Much of the Proprio- and Interoceptor information never rises to consciousness.

Overstimulation of any receptor can give rise to sensation of PAIN.

Most receptors show ADAPTATION — if continuously stimulated they send reduced numbers of impulses to the brain.

PERCEPTION is awareness of the source of the stimulus.
Information from one Sense Organ is correlated with other information (past or present).

APPERCEPTION is recognition or identification of the source of the stimulus and depends on association with past experience.

SMELL

Smell is a CHEMICAL SENSE, i.e. the receptors respond to CHEMICAL STIMULI. To arouse the sensation a substance must first be in a GASEOUS STATE then go into SOLUTION.

The ORGAN of SMELL is the NOSE —— *Also serves as the main air passage to Respiratory System.*

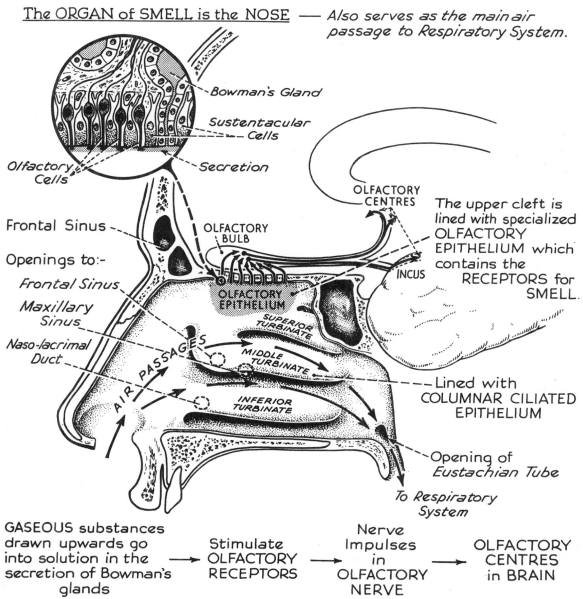

Bowman's Gland

Sustentacular Cells

Secretion

Olfactory Cells

OLFACTORY CENTRES

The upper cleft is lined with specialized OLFACTORY EPITHELIUM which contains the RECEPTORS for SMELL.

Frontal Sinus

Openings to:-

Frontal Sinus

Maxillary Sinus

Naso-lacrimal Duct

OLFACTORY BULB

OLFACTORY EPITHELIUM

SUPERIOR TURBINATE

MIDDLE TURBINATE

INFERIOR TURBINATE

AIR PASSAGES

INCUS

Lined with COLUMNAR CILIATED EPITHELIUM

Opening of Eustachian Tube

To Respiratory System

GASEOUS substances drawn upwards go into solution in the secretion of Bowman's glands → Stimulate OLFACTORY RECEPTORS → Nerve Impulses in OLFACTORY NERVE → OLFACTORY CENTRES in BRAIN

TASTE

Taste is a CHEMICAL SENSE, i.e. receptors respond to CHEMICAL STIMULI. To arouse the sensation a substance must be in SOLUTION.

The essential <u>ORGAN of TASTE</u> is the TONGUE ——— The VOLUNTARY muscular organ concerned also in MASTICATION, SWALLOWING and SPEECH.

Covered with *STRATIFIED SQUAMOUS EPITHELIUM*.

Projections on its upper surface are called

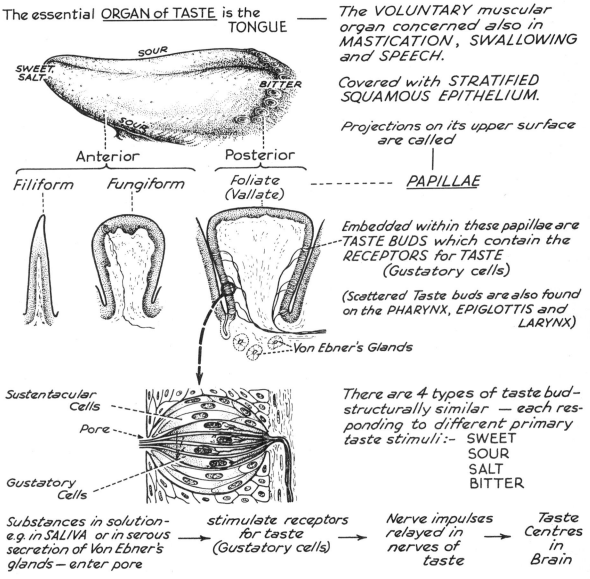

PAPILLAE

Embedded within these papillae are TASTE BUDS which contain the RECEPTORS for TASTE (Gustatory cells)

(Scattered Taste buds are also found on the PHARYNX, EPIGLOTTIS and LARYNX)

Von Ebner's Glands

There are 4 types of taste bud— structurally similar — each responding to different primary taste stimuli:— SWEET
SOUR
SALT
BITTER

Substances in solution— e.g. in SALIVA or in serous secretion of Von Ebner's glands — enter pore → stimulate receptors for taste (Gustatory cells) → Nerve impulses relayed in nerves of taste → Taste Centres in Brain

Other tastes are probably due to combinations of these with smell or with ordinary skin sensations.

PATHWAYS and CENTRES for TASTE

The RECEPTORS for TASTE are linked by a chain of 3 Neurones with the RECEIVING CENTRES for TASTE in the CEREBRAL CORTEX.

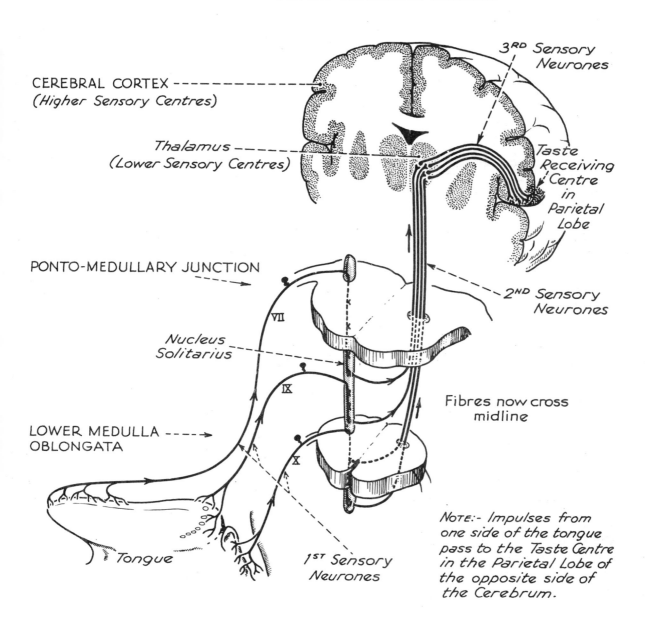

3RD Sensory Neurones

CEREBRAL CORTEX
(Higher Sensory Centres)

Thalamus
(Lower Sensory Centres)

Taste Receiving Centre in Parietal Lobe

PONTO-MEDULLARY JUNCTION

2ND Sensory Neurones

VII

Nucleus Solitarius

IX

LOWER MEDULLA OBLONGATA

Fibres now cross midline

X

Tongue

1ST Sensory Neurones

NOTE:- Impulses from one side of the tongue pass to the Taste Centre in the Parietal Lobe of the opposite side of the Cerebrum.

223

EYE

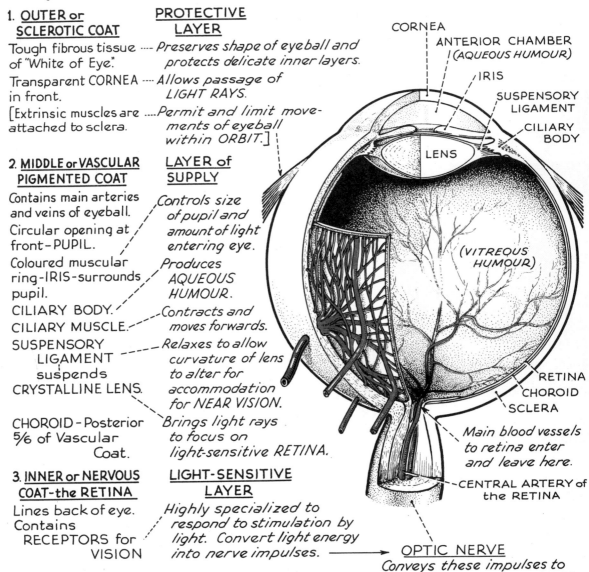

STRUCTURE | FUNCTION of PARTS

The Eyeball has three coats :-

1. OUTER or SCLEROTIC COAT — PROTECTIVE LAYER

Tough fibrous tissue .... *Preserves shape of eyeball and*
of "White of Eye". *protects delicate inner layers.*

Transparent CORNEA .... *Allows passage of*
in front. *LIGHT RAYS.*

[Extrinsic muscles are ....*Permit and limit move-*
attached to sclera. *ments of eyeball*
within ORBIT.]

2. MIDDLE or VASCULAR PIGMENTED COAT — LAYER of SUPPLY

Contains main arteries
and veins of eyeball.

Circular opening at *Controls size*
front – PUPIL. *of pupil and*
amount of light
Coloured muscular *entering eye.*
ring – IRIS – surrounds
pupil. *Produces*
CILIARY BODY. *AQUEOUS*
HUMOUR.
CILIARY MUSCLE. *Contracts and*
moves forwards.
SUSPENSORY *Relaxes to allow*
LIGAMENT *curvature of lens*
suspends *to alter for*
CRYSTALLINE LENS. *accommodation*
for NEAR VISION.

CHOROID – Posterior *Brings light rays*
5/6 of Vascular *to focus on*
Coat. *light-sensitive RETINA.*

3. INNER or NERVOUS COAT – the RETINA — LIGHT-SENSITIVE LAYER

Lines back of eye. *Highly specialized to*
Contains *respond to stimulation by*
RECEPTORS for *light. Convert light energy*
VISION *into nerve impulses.* ⟶

CORNEA
ANTERIOR CHAMBER (AQUEOUS HUMOUR)
IRIS
SUSPENSORY LIGAMENT
CILIARY BODY
LENS
(VITREOUS HUMOUR)
RETINA
CHOROID
SCLERA
Main blood vessels to retina enter and leave here.
CENTRAL ARTERY of the RETINA

OPTIC NERVE
Conveys these impulses to
VISUAL CENTRES in OCCIPITAL
(posterior) part of BRAIN.

PROTECTION of the EYE

The hidden posterior 4/5 of the eyeball is encased in a bony socket – the ORBITAL CAVITY.   A thick layer of *Areolar* and *Adipose* tissue forms a cushion between bone and eyeball.   The exposed anterior 1/5 of the eyeball is protected from injury by:-

The <u>EYELIDS</u> ------------------- close reflexly to protect eye from dust
     Fringed with Eyelashes   *and other foreign particles.*

<u>CONJUNCTIVA</u> ----------------- *smooth surfaces which glide over each*
   A delicate membrane lining eyelids   *other when lids open and close.*
and covering exposed surface of eye.

<u>LACRIMAL GLANDS</u> -------------- *continuously secrete TEARS. These*
                              flow over, wash and lubricate surface
                              of eye. They contain an ENZYME –
                              LYSOZYME – which destroys bacteria.

<u>TARSAL GLANDS</u> ----------------------- *secrete a fluid to prevent lids*
                                      from sticking together

<u>LACRIMAL DUCTS</u>
   drain tears
   from surface
   of eye

--- CONJUNCTIVA

--- LACRIMAL GLANDS

<u>LACRIMAL SAC</u>

--- OPENINGS OF
     TARSAL GLANDS

ORBICULARIS OCULI MUSCLE

<u>NASO-LACRIMAL</u>
<u>DUCT</u> ------------------- *drains tears into back of nose.*

MUSCLES of EYE

The EYEBALLS are moved by SMALL MUSCLES which link the SCLEROTIC COAT to the BONY SOCKET.

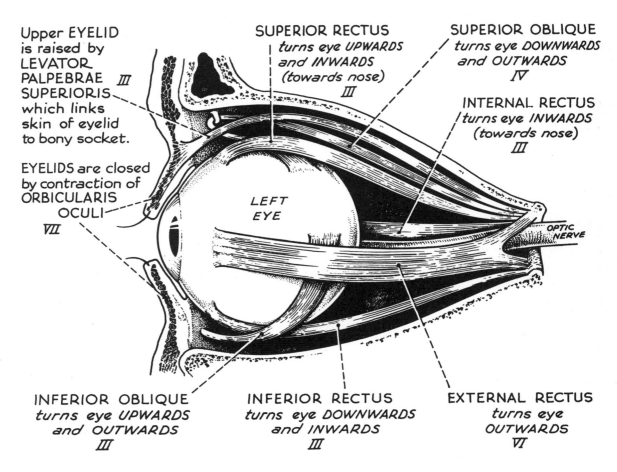

Upper EYELID is raised by LEVATOR PALPEBRAE SUPERIORIS *III* which links skin of eyelid to bony socket.

EYELIDS are closed by contraction of ORBICULARIS OCULI *VII*

SUPERIOR RECTUS
turns eye UPWARDS and INWARDS (towards nose) III

SUPERIOR OBLIQUE
turns eye DOWNWARDS and OUTWARDS IV

INTERNAL RECTUS
turns eye INWARDS (towards nose) III

LEFT EYE

OPTIC NERVE

INFERIOR OBLIQUE
turns eye UPWARDS and OUTWARDS III

INFERIOR RECTUS
turns eye DOWNWARDS and INWARDS III

EXTERNAL RECTUS
turns eye OUTWARDS VI

Acting together, the Extrinsic muscles of the Eyeballs can bring about ROTATORY movements of the Eyes.
The Extrinsic muscles are supplied by motor fibres from Cranial Nerves III, IV and VI.

226

CONTROL of EYE MOVEMENTS

Both eyes normally move together so that images continue to fall on corresponding points of both retinae.

VOLUNTARY EYE MOVEMENTS
are initiated in motor centres in FRONTAL LOBES.

Impulses from <u>one side</u> of the Cerebral Cortex turn <u>both eyes</u> to the <u>other side</u> of Visual Field.

REFLEX EYE MOVEMENTS
Two Groups – (1) Those in response to Visual Stimuli. (2) Those in response to Non-Visual Stimuli. In control of these are :- CENTRES in OCCIPITAL LOBES: CENTRES in MIDBRAIN and PONS which give rise to CRANIAL NERVES III, IV and VI.

Impulses from <u>one side</u> of the Midbrain and Pons turn eyes to the <u>same side</u>.

These centres are <u>closely linked</u> with each other and with HIGHER and LOWER centres in the Central Nervous System, so that the eyes are moved reflexly in response to many stimuli, e.g. loud noises or proprioceptive messages from vestibular organs.

e.g.
From the <u>RIGHT</u> side of the midbrain and pons impulses pass to the the <u>PRIME MOVERS</u>– RIGHT EXT. RECTUS & LEFT INT. RECTUS – which contract (+).
The <u>ANTAGONISTS</u>– LEFT EXT. RECTUS & RIGHT INT. RECTUS – relax (–) to permit the movement.
<u>BOTH EYES</u> turn to the <u>RIGHT</u>.

This is <u>CONJUGATE DEVIATION</u>

III

MIDBRAIN

IV

MEDIAL LONGITUDINAL BUNDLES

VI

PONS

IRIS, LENS and CILIARY BODY

The **IRIS** is a muscular diaphragm with a central opening - the PUPIL.

The **LENS** is a transparent biconvex crystalline disc.

IRIS controls amount of LIGHT entering the EYE

CIRCULAR smooth muscle fibres –*SPHINCTER PUPILLAE*– contract to make Pupil smaller in bright light.

RADIAL fibres—
        DILATOR PUPILLAE– contract to make Pupil larger with change from LIGHT to DARK; NEAR to DISTANT vision (also with FEAR and PAIN).

When CILIARY MUSCLE contracts, SUSPENSORY LIGAMENT is slackened. Tension on CAPSULE of LENS is relaxed. Anterior surface bulges forwards ⟶ LENS becomes more convex especially in its central part. This brings near objects into focus. (Accommodation Reflex)

Outer elastic CAPSULE blends with SUSPENSORY LIGAMENT which suspends LENS behind IRIS

The LENS and IRIS are attached to **CILIARY BODY** which contains fibres of SMOOTH MUSCLE

LENS brings LIGHT RAYS to a FOCUS upside down on the RETINA. To do this under all conditions it must be able to alter its curvature.

These changes are brought about reflexly. The ingoing impulses travel in the Optic nerves. The outgoing motor impulses travel in Parasympathetic to Ciliary Body and Sphincter Pupillae and in Sympathetic to Dilator Pupillae.

ACTION of LENS

The normal lens brings light rays to a sharp focus upside down on the retina. It can do this whether we are looking at an object far away or one close at hand. The curvature increases reflexly to accommodate for near vision.

Rays of light coming from every point of a DISTANT object (over 20 feet away) are PARALLEL.

They pass through the CORNEA, AQUEOUS HUMOUR

and LENS which refract them to a sharp focus — upside down and reversed from side to side – on the Retina.

The conscious mind learns to interpret the image and project it to its true position in space.

Rays of light coming from a NEAR object (less than 20 feet away) RADIATE from every point

A more convex lens is required to bring these rays to a sharp focus on the Retina.

If the EYEBALL is _too short_, rays from a distant object are brought into focus BEHIND the Retina when the ciliary muscle is relaxed.

Vision is blurred.

This is longsightedness or HYPERMETROPIA.

The longsighted eye has to accommodate even for distant vision: i.e. Ciliary muscles contract to give a more convex lens and distant objects are then seen clearly. This limits amount of accommodating power left for near objects and the nearest point for sharp vision is then further away. It can be corrected by fitting spectacles with an additional convex lens.

If the EYEBALL is _too long_, rays from a distant object are brought into focus IN FRONT of the Retina.

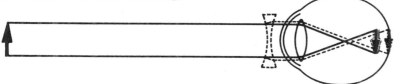

This is shortsightedness or MYOPIA — only objects near the eye can be seen clearly.

It can be corrected by using a concave lens.

FUNDUS OCULI

Part of the RETINA can be seen by means of an instrument – the OPHTHALMOSCOPE – which shines a beam of light through the PUPIL of the eye on to the RETINA.

The part of the Retina seen in this way is called the FUNDUS OCULI.

OPTIC PAPILLA – or 'BLIND SPOT'– The nerve fibres from all parts of the retina converge on this area to leave the eyeball as the OPTIC NERVE. It has no RODS or CONES and therefore is not itself sensitive to light.
RETINAL BLOOD VESSELS enter or leave the eyeball here.

MACULA LUTEA – or 'YELLOW SPOT' – with
FOVEA CENTRALIS – area of acute vision – contains CONES only (the Receptors stimulated in BRIGHT and COLOURED LIGHT). When we look at an object the eyes are directed so that the image will fall on the fovea of each eye.

EXTRAFOVEAL part of Retina – area of less acute vision – CONES become fewer – RODS (the Receptors stimulated in DIM LIGHT with no discrimination between COLOURS) become more numerous towards the peripheral part of the Retina.

230

RETINA

Sections of the Retina examined under the microscope show 8 layers :-

**LAYER of
PIGMENT CELLS** next to CHOROID COAT.

> *In <u>bright</u> light pigment granules migrate into the cell processes lying between rods and cones. This prevents spread of light from one receptor to another. In <u>dim</u> light the granules are confined to the cell body.*

RODS and CONES — *When light strikes these*
- 1ST (Receptive) Neurones

Receptors for Vision, impulses are set up which are transmitted via

OUTER NUCLEAR LAYER
- Nuclei of Rods and Cones

OUTER PLEXIFORM LAYER
- Nerve Processes and Synapses between Neurones

INNER NUCLEAR LAYER
- Cell Bodies of 2ND or INTERMEDIATE NEURONES

INNER PLEXIFORM LAYER
- Nerve Processes and Synapses between Neurones

LAYER of GANGLION CELLS
- Cell Bodies of 3RD or INTEGRATING NEURONES

LAYER of NERVE FIBRES
- These converge on OPTIC PAPILLA to leave Eyeball as the OPTIC NERVE

to
Visual Area of
CEREBRAL CORTEX

FOVEA CENTRALIS contains only CONES. All other layers are thin in this part of RETINA.

Note:- LIGHT rays must pass through all these layers except the pigment cell layer to reach and stimulate RECEPTORS.

231

MECHANISM of VISION

WHITE LIGHT is really due to the fusion of *coloured lights*. These coloured lights are separated by shining a beam of white light through a glass prism. This is called the VISIBLE SPECTRUM.

7,800 Å	7,000 Å	6,000 Å			5,000 Å			3,900 Å
RED		ORANGE	YELLOW	YELL.-GR.	GREEN	BLUE-GREEN	BLUE	VIOLET

If light is bright or intense the spectrum appears brightest to man's eye in the orange band (6,100 Å).

If brightness or intensity of light source is gradually reduced, colour perception is gradually lost and the spectrum appears as a luminous band with a very dark area in the red band and with its brightest part in the green band (5,300Å).

Visual Receptors contain PIGMENTS which break down chemically in the presence of specific wavelengths of light. This forms a chemical stimulus which in some way triggers off nerve impulses which travel from the retina to the cerebral cortex.

<u>SCOTOPIC VISION</u> is vision in <u>DIM LIGHT</u>.   It depends on the <u>RODS</u>

<u>RODS</u> are of one type ———————and give——————<u>MONOCHROMATIC VISION</u>

RODS contain   RHODOPSIN or SCOTOPSIN } ("VISUAL PURPLE")

In DIM LIGHT
absorbs light / *Pigment gradually splits* → RETINENE + OPSIN *(a protein)*

[In BRIGHT LIGHT
RHODOPSIN - very rapid breakdown with "bleaching".*]*

Chemical Stimulus triggers off <u>*Nerve Impulse*</u>

In DARKNESS
RHODOPSIN ←—— *regeneration* ——— RETINENE + OPSIN
(regenerated from Vit.A)

When a person passes from a brightly lit scene to darkness he is temporarily blinded. After about ½ hour he sees well.   This adjustment or increase in sensitivity is called <u>DARK ADAPTATION</u>.

As light brightness or intensity increases rods lose their sensitivity and cease to respond.

MECHANISM of VISION

PHOTOPIC VISION is vision in BRIGHT LIGHT. It depends on the CONES

CONES are thought to be of 3 types —— giving TRICHROMATIC VISION.
      Each type with a
        different
     photosensitive
       VISUAL
       PIGMENT with its own wavelength to which it is sensitive, which
                it absorbs and by which it is broken down to form the
                                           chemical stimulus.

Note:- The Photosensitive pigments in cones have not yet been isolated.

     "RED" receptors absorb YELLOW—ORANGE light
    "GREEN" receptors absorb   GREEN   light
    "BLUE" receptors absorb   BLUE   light

All 3 types of "PHOTOPSINS" are probably stimulated in roughly equal
                     proportions when WHITE light falls on retina:
2 or more in varying proportions when OTHER COLOURS of light fall on
                                         retina.

The various types of colour blindness could be explained in terms of the absence or deficiency of one or more of these special receptors.

[*A colour sensation has 3 qualities:-*

   HUE ———————— depends largely on wavelength.

   SATURATION —— purity —
           A "saturated" colour has no white light mixed with it.
           An "unsaturated" colour has some white light mixed with it.

   INTENSITY —— brightness — depends largely on "strength" of the light.]

As the intensity of light is reduced the Cones cease to respond and the Rods take over.

When a person passes from darkness to bright light he is dazzled but after a short time he sees well again.
    This adjustment or decrease in sensitivity on exposure to bright light is called LIGHT ADAPTATION.

VISUAL PATHWAYS to the BRAIN

The RECEPTORS for VISION are linked by a chain of Neurones with RECEIVING and INTEGRATING CENTRES in the OCCIPITAL LOBES of the CEREBRAL CORTEX.

The two VISUAL FIELDS are slightly DISSIMILAR but OVERLAP.

LIGHT RAYS from LEFT side of Visual fields fall on the
INNER (Nasal) HALF of LEFT RETINA
and
OUTER (Temporal) HALF of RIGHT RETINA.

Impulses are relayed from RECEPTOR Neurones by INTERMEDIATE to INTEGRATING Neurones whose axons form the OPTIC NERVES.

Fibres from the NASAL HALF of each RETINA cross in OPTIC CHIASMA.

Fibres from the TEMPORAL HALF of each RETINA remain uncrossed.

Most fibres SYNAPSE with Neurones in the LOWER VISUAL CENTRES – LATERAL GENICULATE BODY

[Some fibres synapse in Pretectal Region, some synapse in Superior Colliculus — these form afferents for the OPTIC REFLEXES.]

The axons of these Neurones form OPTIC RADIATIONS
They relay VISUAL IMPULSES to HIGHER VISUAL CENTRES in OCCIPITAL LOBES where the images from both eyes are integrated into a single SENSATION.

MIDBRAIN

Note:- One side of the OCCIPITAL CORTEX receives impressions from the FIELD of VISION on the opposite side.

STEREOSCOPIC VISION

When we look at some object or scene the view seen by the RIGHT EYE is slightly different from the view seen by the LEFT EYE.

These two DISSIMILAR RETINAL IMAGES are fused in the Visual Centres of the Brain to give a 3-dimensional picture – an appreciation of DEPTH as well as of HEIGHT and WIDTH.

Other factors contribute to our recognition of SOLIDITY, DEPTH and DISTANCE :-

COLOURS FADE with distance and DETAILS become indistinct.

LIGHT and SHADE on surface of object.

SIZE of IMAGES cast on RETINA may be small when actual size is known to be large.

OCCLUSION or blocking out of parts of a distant object by things between it and the eyes.

PERSPECTIVE - straight lines known to be parallel appear to converge with distance.

PARALLAX - when head is moved from side to side objects near at hand appear to move in opposite direction while those farther away may seem to move with the eyes.

OCCIPITAL LOBE

By complex mental processes these points are interpreted in terms of Distance and Depth.

LIGHT REFLEX

When LIGHT falls on the RETINA the PUPILS CONTRACT.

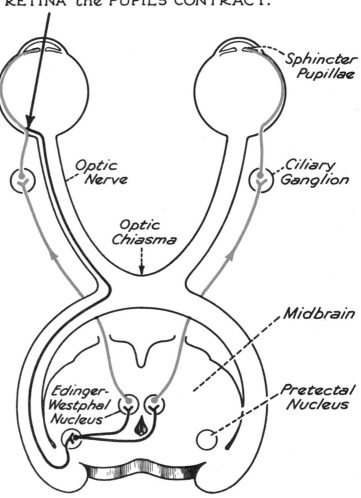

Impulses from
the Receptors in
the Retina pass
in the
OPTIC NERVE
to the
PRETECTAL
NUCLEUS
in the midbrain
of the same side

Impulses then
pass to the
EDINGER-WESTPHAL
NUCLEUS
of the 3ᴿᴰ Cranial
Nerves of <u>both sides</u>

Impulses now travel
in efferent
Parasympathetic motor
nerves to cause
contraction of both
SPHINCTER
PUPILLAE

Sphincter Pupillae

Optic Nerve

Ciliary Ganglion

Optic Chiasma

Midbrain

Edinger-Westphal Nucleus

Pretectal Nucleus

DIRECT LIGHT REFLEX

*When light shines into
one eye (as in diagram)
the PUPIL of that eye contracts* ———

CONSENSUAL LIGHT REFLEX

*the PUPIL of the other eye also
contracts.*

*This cuts down the amount of light entering the eyes and protects
the Retinae from excessive stimulation. It also increases depth of
focus and improves the sharpness of the Retinal images.*

EAR

The ear has 3 separate parts, each with different rôles in the mechanism of HEARING:-

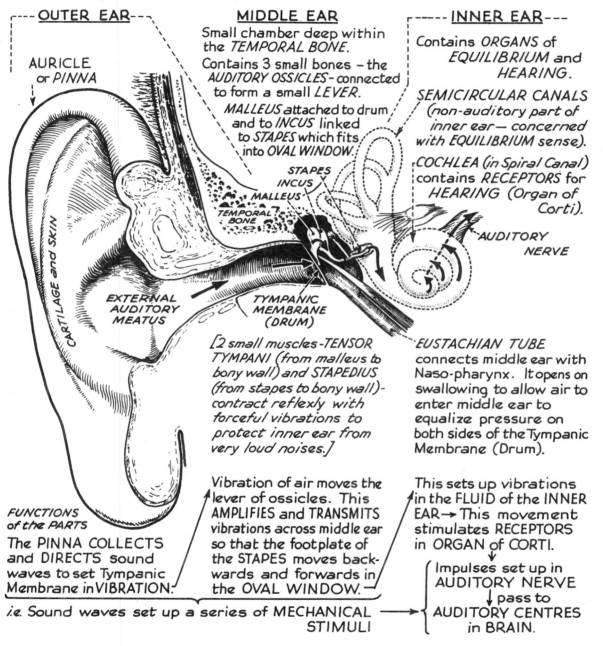

---- **OUTER EAR** ---

AURICLE or *PINNA*

CARTILAGE and SKIN

EXTERNAL AUDITORY MEATUS

MIDDLE EAR

Small chamber deep within the *TEMPORAL BONE*. Contains 3 small bones – the *AUDITORY OSSICLES* - connected to form a small *LEVER*.

MALLEUS attached to drum and to *INCUS* linked to *STAPES* which fits into *OVAL WINDOW*.

STAPES
INCUS
MALLEUS
TEMPORAL BONE

TYMPANIC MEMBRANE (DRUM)

----- **INNER EAR** ---

Contains *ORGANS* of *EQUILIBRIUM* and *HEARING*.

SEMICIRCULAR CANALS (non-auditory part of inner ear — concerned with EQUILIBRIUM sense).

COCHLEA (in Spiral Canal) contains *RECEPTORS* for *HEARING (Organ of Corti)*.

AUDITORY NERVE

[2 small muscles -TENSOR TYMPANI (from malleus to bony wall) and STAPEDIUS (from stapes to bony wall)- contract reflexly with forceful vibrations to protect inner ear from very loud noises.]

EUSTACHIAN TUBE connects middle ear with Naso-pharynx. It opens on swallowing to allow air to enter middle ear to equalize pressure on both sides of the Tympanic Membrane (Drum).

FUNCTIONS of the PARTS

The PINNA COLLECTS and DIRECTS sound waves to set Tympanic Membrane in VIBRATION.

Vibration of air moves the lever of ossicles. This AMPLIFIES and TRANSMITS vibrations across middle ear so that the footplate of the STAPES moves backwards and forwards in the OVAL WINDOW.

This sets up vibrations in the FLUID of the INNER EAR → This movement stimulates RECEPTORS in ORGAN of CORTI.

Impulses set up in AUDITORY NERVE pass to AUDITORY CENTRES in BRAIN.

i.e. Sound waves set up a series of MECHANICAL STIMULI ⟶

COCHLEA

The Cochlea is the essential organ of HEARING.

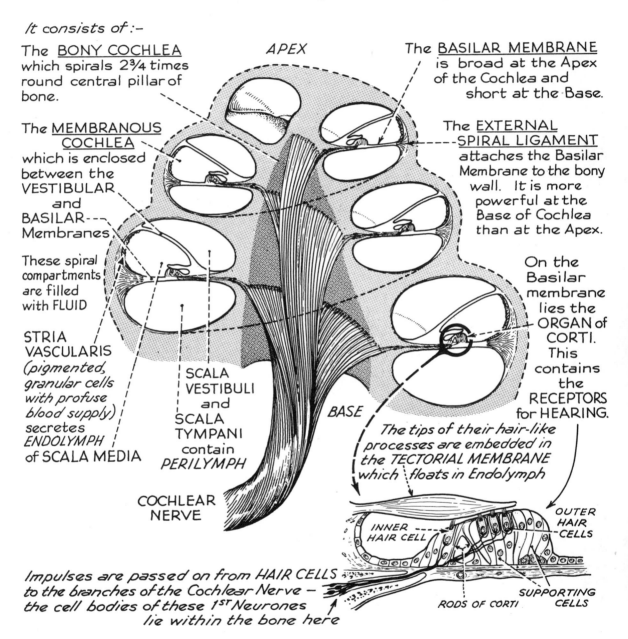

It consists of :-

The BONY COCHLEA which spirals 2¾ times round central pillar of bone.

The MEMBRANOUS COCHLEA which is enclosed between the VESTIBULAR and BASILAR Membranes

These spiral compartments are filled with FLUID

STRIA VASCULARIS *(pigmented, granular cells with profuse blood supply)* secretes *ENDOLYMPH* of SCALA MEDIA

APEX

SCALA VESTIBULI and SCALA TYMPANI contain *PERILYMPH*

COCHLEAR NERVE

BASE

The BASILAR MEMBRANE is broad at the Apex of the Cochlea and short at the Base.

The EXTERNAL SPIRAL LIGAMENT attaches the Basilar Membrane to the bony wall. It is more powerful at the Base of Cochlea than at the Apex.

On the Basilar membrane lies the ORGAN of CORTI. This contains the RECEPTORS for HEARING.

The tips of their hair-like processes are embedded in the TECTORIAL MEMBRANE which floats in Endolymph

INNER HAIR CELL

OUTER HAIR CELLS

RODS OF CORTI

SUPPORTING CELLS

Impulses are passed on from HAIR CELLS to the branches of the Cochlear Nerve — the cell bodies of these 1ST Neurones lie within the bone here

MECHANISM of HEARING

This is most readily understood if the COCHLEA is imagined as straightened out:—

VIBRATIONS of AIR - - → set in motion DRUM and OSSICLES

BASE ——————————— *APEX*

STAPES plunges in and out of OVAL WINDOW. Fluid is set vibrating in SCALA VESTIBULI

These agitations are transmitted to the thin-walled SCALA MEDIA *(DUCTUS COCHLEARIS)*

SCALA VESTIBULI
VESTIBULAR MEMBRANE
SCALA MEDIA
ORGAN OF CORTI
BASILAR MEMBRANE
SCALA TYMPANI

ROUND WINDOW

Travelling waves are set up at the Basal end of the BASILAR MEMBRANE. Distance travelled depends on the frequency of the note.

e.g. High frequency waves reach maximum amplitude soon, then die out.

Low frequency waves travel further before reaching their peak.

Only the hair cells over-lying the point of greatest movement are stimulated

On this theory the appreciation of the "Pitch" of a note is the function of the COCHLEA.
Notes of frequencies ranging from 20 to 20,000 double vibrations (cycles) per second can be heard by man.

BASE	*1ST Turn*	*2ND Turn*	*3RD Turn*	APEX
		BASILAR MEMBRANE		
	1,500 to 20,000 cycles per second	500 to 1,500 cycles per second	20 to 500 cycles per second	

AUDITORY PATHWAYS to BRAIN

The RECEPTORS for HEARING are linked by a chain of NEURONES with the RECEIVING CENTRES for HEARING in the TEMPORAL LOBES of the CEREBRAL CORTEX.

CEREBRAL CORTEX

Auditory Receiving
Centres
in
TEMPORAL LOBE

MIDBRAIN

Cochlear] Dorsal
Nuclei]Ventral

UPPER MEDULLA
OBLONGATA

Medial Geniculate
Body

Inferior Quadrigeminal
Body

Lateral Lemniscus

Olivary
Nucleus

Trapezoid
Body

[Many fibres cross
over; some travel
up on same side]

Auditory
Nerve

Sound
waves stimulate Receptor Cells in
Organ of Corti

Impulses travel in Auditory Nerve and
are relayed as shown.

240

SPECIAL PROPRIOCEPTORS

The "Special" Proprioceptors of the body are found in the Non-Auditory part of the Inner Ear — the Labyrinth. They are stimulated by movements or change of position of the head in space.

Three **SEMICIRCULAR CANALS** — *in each inner ear – one in each of three planes of space* – contain Receptors.

Membranous tubes containing Endolymph

These receptors, situated in the Ampulla of each canal, are stimulated mechanically by the *starting or stopping of rotatory movements* of the head in space.

One end of each canal has a swelling - the AMPULLA

Both ends of each canal open into the UTRICLE

In the **UTRICLE** and **SACCULE** a Macula contains Receptors which are stimulated by *changes in position* of head (e.g. tilting)

OTOLITHS *(crystals of calcium carbonate)* in a GELATINOUS mass mechanically stimulate the HAIR PROCESSES of HAIR CELLS
↓
Information carried in fibres of Vestibular Nerve to Centres in Brain.

A GELATINOUS mass – the Cupula — stretches across each Ampulla. Its apex swings free and is moved by movements of the endolymph.
↓
HAIR PROCESSES are bent
↓
HAIR CELLS are stimulated
↓
Messages are carried by Vestibular branch of VIII Cr. N. to Centres in Brain.

ORGAN of EQUILIBRIUM: MECHANISM of ACTION

<u>*At Rest*</u>

<u>RIGHT</u>

Lateral
Horizontal
Canal

AMPULLA
CUPULA
CRISTA
BRANCHES
of VESTIBULAR
NERVE

<u>LEFT</u>

Lateral
Horizontal
Canal

<u>*When head starts to rotate*</u> (e.g. to the right) the endolymph in the semicircular canals which lie at right angles to the axis of rotation tends to lag behind the movement of the head — and the

CUPULA is displaced,
HAIR CELLS are stimulated and
INGOING impulses form afferent pathways
for Reflexes leading to alterations in tone
of muscles in neck, trunk and limbs to
avoid body losing balance.

<u>*After the initial inertia is overcome*</u> the endolymph no longer lags behind the movement of the head — and the

CUPULA is no longer displaced.
HAIR CELLS are no longer bent and
stimulated.
Nerve fibres no longer send signals
to medulla and cerebellum.

<u>*When head stops rotating*</u> the endolymph tends to continue to rotate and the

CUPULA is displaced in the opposite
direction → HAIR CELLS are bent →
Nerve fibres signal rotation of head
to left → Individual feels for a
moment as though he is rotating
in opposite direction — when in fact
he has ceased to rotate.

Rotating the head round a horizontal axis — i.e. tilting it backwards and forwards such as happens in the pitch and roll of a ship — stimulates the vertical canals. This can lead to "motion sickness".

Stimulation of the Semicircular canals also causes movements of the eyes to avoid too much displacement of the image being cast on the retinae.

During rotation there is a slow movement of the eyes in the direction opposite to that of rotation, then a quick return to the normal position. This is NYSTAGMUS. It continues for a short time after movement has ceased.

242

VESTIBULAR PATHWAYS to BRAIN

The Special Proprioceptive end-organs in the LABYRINTH are linked through the VESTIBULAR NUCLEI with RECEIVING and INTEGRATING CENTRES in the CEREBELLUM, and with MOTOR CENTRES in the MIDBRAIN and SPINAL CORD through which they initiate reflex muscular movements of EYES, HEAD and NECK, TRUNK and LIMB muscles to adjust balance and posture.

CEREBRAL CORTEX

MIDBRAIN

CEREBELLUM

Vestibular Nuclei { Superior
Medial
Lateral

UPPER MEDULLA OBLONGATA

Vestibular Nerve Ganglion

Semicircular Canals

Lateral Vestibulo-spinal Tract

Medial Geniculate Body

To Nuclei for Eye muscles

Medial Longitudinal Bundles

SPINAL CORD

GENERAL PROPRIOCEPTORS

Proprioceptors are the sense organs stimulated by MOVEMENT of the body itself. They make us aware of the movement or position of the body in space and of the various parts of the body to each other. They are important as ingoing afferent pathways in reflexes for adjusting Posture and Tone.

General Proprioceptors are found in SKELETAL MUSCLES, TENDONS and JOINTS.

GOLGI ORGAN — in tendons —
stimulated by tension which occurs when muscle is STRETCHED and when it is CONTRACTED

MUSCLE SPINDLE -
—in skeletal muscle-
stimulated when muscle is STRETCHED or SHORTENED

PACINIAN CORPUSCLES
similar to those in the skin are found in Deep Connective Tissue and around Joints. They are stimulated by PRESSURE of surrounding structures when joints are moved.

Muscle spindles themselves contain specialized muscle fibres. These intra-fusal fibres are supplied with fine motor nerves which are probably under the control of the Cerebellum (see page 260)

PROPRIOCEPTOR PATHWAYS to BRAIN

General Proprioceptor end-organs may be linked with centres in
(a) CEREBELLUM or (b) PARIETAL LOBE of CEREBRAL CORTEX by a chain
of 3 Neurones.

CEREBRAL CORTEX ------

Thalamus ----

3RD Sensory Neurone
passes from Thalamus
→ Internal Capsule →
to reach Post Central
Gyrus of Parietal
Lobe – where awareness
of muscle and joint
sense is appreciated.

These fibres
in the CEREBELLUM
link up with others
concerned in maintenance
of posture and sense of
balance or equilibrium.

Proprioceptor impulses
from Head and Neck muscles
travel in Cranial Nerves ------
e.g. from muscles of mastication

V Cranial N.
Sensory   Motor
Nucleus   Nuc.

PONS

Medial Lemniscus

Gracile Nucleus
(fibres from lower limbs
and lower trunk)

LOWER MEDULLA
OBLONGATA ------

Cuneate Nucleus
(fibres from upper
limbs and trunk)

2ND Sensory Neurone
crosses over to
opposite side of
Brain Stem in
Sensory Decussation

1ST Sensory
Neurones

Dorsal Spino-cerebellar Tract

SPINAL CORD

Proprioceptor impulses
travel from Trunk and Limbs in Spinal Nerves.

CUTANEOUS SENSATION

There are 5 basic skin sensations — TOUCH, PRESSURE, PAIN, WARMTH and COLD. There is much controversy as to how these are registered. In some areas they appear to be served by special nerve endings (sensory receptors or end-organs) in the SKIN. These end-organs are not uniformly distributed over the whole body surface. (E.g. Touch "endings" are very numerous in hands and feet but are much less frequent in the skin of the back.)

In *HAIRLESS* parts of the skin (the palms of the hands and soles of the feet)

MEISSNER corpuscles register TOUCH.

KRAUSE bulbs (*function uncertain*).

RUFFINI endings (*function uncertain*).

PACINIAN corpuscles deep in dermis are stimulated by PRESSURE.

In *HAIRY* surfaces of body

Network of nerve fibres round SHAFT of HAIR registers sensations of TOUCH when hair is moved.

Branching nerve endings in epidermis and dermis register PAIN and probably other skin sensations.

Tickling, itching, softness, hardness, wetness are probably due to stimulation of two or more of these special endings and to a blending of the sensations in the Brain.

246

SENSORY PATHWAYS from SKIN of FACE

The Receptors or nerve endings for ORDINARY SKIN SENSATIONS are linked by 3 Neurones with Receiving Centres in the PARIETAL LOBES.

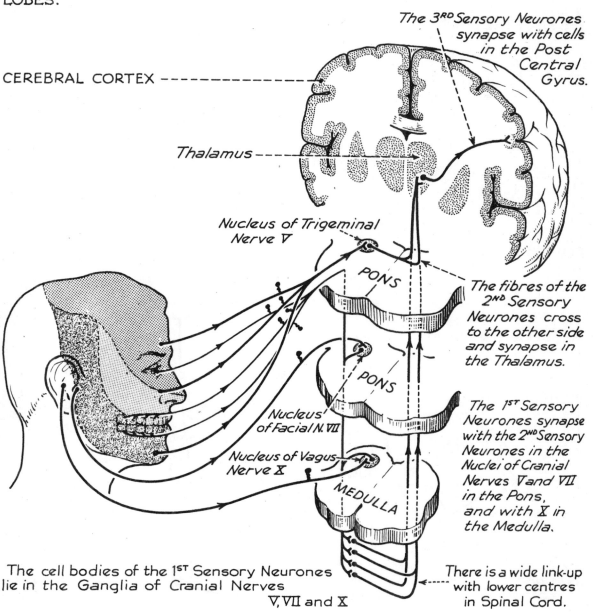

The 3RD Sensory Neurones synapse with cells in the Post Central Gyrus.

CEREBRAL CORTEX

Thalamus

Nucleus of Trigeminal Nerve V

PONS

The fibres of the 2ND Sensory Neurones cross to the other side and synapse in the Thalamus.

PONS

Nucleus of Facial N. VII

The 1ST Sensory Neurones synapse with the 2ND Sensory Neurones in the Nuclei of Cranial Nerves V and VII in the Pons, and with X in the Medulla.

Nucleus of Vagus Nerve X

MEDULLA

The cell bodies of the 1ST Sensory Neurones lie in the Ganglia of Cranial Nerves V, VII and X

There is a wide link-up with lower centres in Spinal Cord.

PAIN and TEMPERATURE PATHWAYS from TRUNK and LIMBS

The nerve endings registering Pain and Warmth or Cold are linked by a chain of 3 Neurones with final Receiving Centre – the sensory area in the PARIETAL LOBES of the CEREBRAL CORTEX.

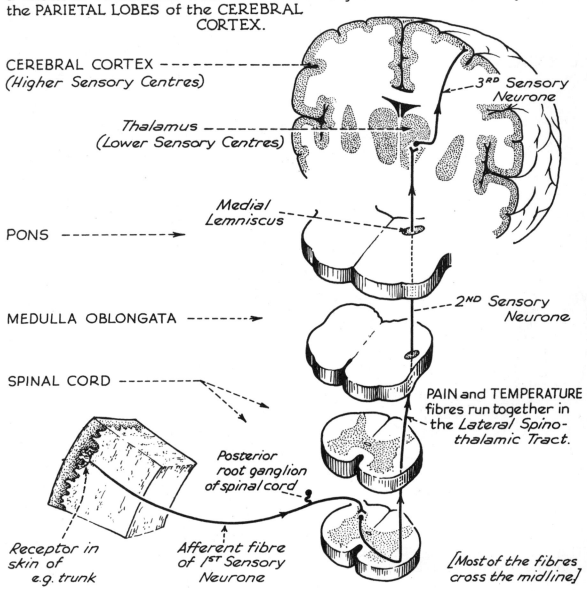

CEREBRAL CORTEX – – – – – – – – –
(Higher Sensory Centres)

3ᴿᴰ Sensory Neurone

Thalamus – – – – – – – –
(Lower Sensory Centres)

Medial Lemniscus – – – – –

PONS – – – – – – – ➤

MEDULLA OBLONGATA – – – – – – ➤

2ᴺᴰ Sensory Neurone

SPINAL CORD – – – – – – – ➤

PAIN and TEMPERATURE fibres run together in the *Lateral Spino-thalamic Tract.*

Posterior root ganglion of spinal cord

Receptor in skin of e.g. trunk

Afferent fibre of 1ˢᵀ Sensory Neurone

[Most of the fibres cross the midline.]

248

TOUCH and PRESSURE PATHWAYS from TRUNK and LIMBS

Touch and Pressure endings are linked by a chain of 3 Neurones with the PARIETAL LOBES.

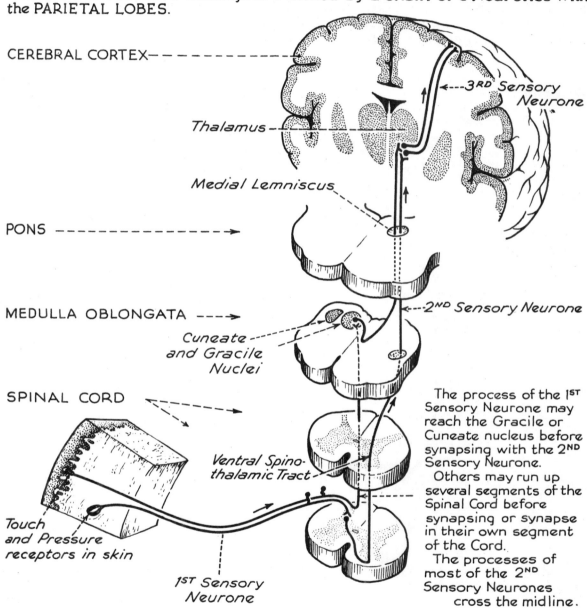

CEREBRAL CORTEX

3RD Sensory Neurone

Thalamus

Medial Lemniscus

PONS

MEDULLA OBLONGATA

2ND Sensory Neurone

Cuneate and Gracile Nuclei

SPINAL CORD

Ventral Spino-thalamic Tract

Touch and Pressure receptors in skin

1ST Sensory Neurone

The process of the 1ST Sensory Neurone may reach the Gracile or Cuneate nucleus before synapsing with the 2ND Sensory Neurone.

Others may run up several segments of the Spinal Cord before synapsing or synapse in their own segment of the Cord.

The processes of most of the 2ND Sensory Neurones cross the midline.

SENSORY CORTEX

The 3ᴿᴰ Sensory Neurones (conveying information from the opposite side of the body) synapse with cells in the POST-CENTRAL GYRUS of the PARIETAL LOBE of the CEREBRAL CORTEX. The exact points on this Gyrus at which impulses coming from the different regions of the skin surface terminate are indicated on this coronal view of the Gyrus

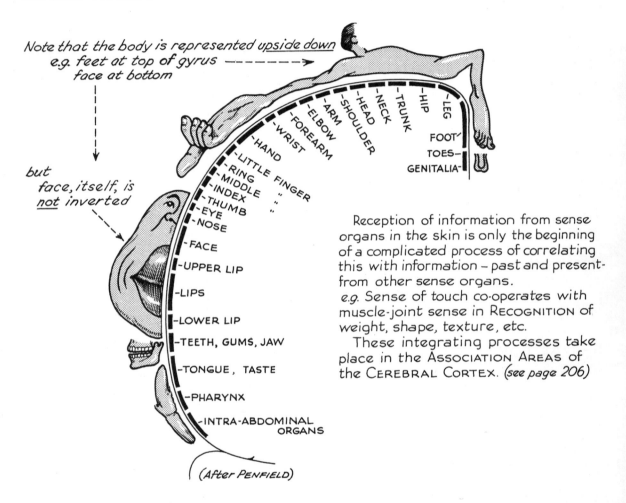

Note that the body is represented upside down
e.g. feet at top of gyrus --------→
face at bottom

but face, itself, is not inverted

LEG
HIP
TRUNK
NECK
HEAD
SHOULDER
ARM
ELBOW
FOREARM
WRIST
HAND
LITTLE FINGER
RING "
MIDDLE "
INDEX "
THUMB "
EYE
NOSE
FACE
UPPER LIP
LIPS
LOWER LIP
TEETH, GUMS, JAW
TONGUE, TASTE
PHARYNX
INTRA-ABDOMINAL ORGANS

FOOT
TOES
GENITALIA

(After PENFIELD)

Reception of information from sense organs in the skin is only the beginning of a complicated process of correlating this with information – past and present– from other sense organs.
e.g. Sense of touch co-operates with muscle-joint sense in RECOGNITION of weight, shape, texture, etc.
These integrating processes take place in the ASSOCIATION AREAS of the CEREBRAL CORTEX. (see page 206)

Note the relatively large area devoted to FACE (especially lips) and to HAND (especially thumb and index finger) while trunk representation is very small.

MOTOR CORTEX

The MOTOR NERVE CELLS which send out impulses to initiate VOLUNTARY MOVEMENT of SKELETAL MUSCLES lie in the PRE-CENTRAL GYRUS of each FRONTAL LOBE in the CEREBRAL CORTEX.

One cerebral hemisphere controls the muscles on the opposite side of the body.

The exact point in the GYRUS where neurones controlling any one part of the body are situated is indicated in this coronal view of the Gyrus.

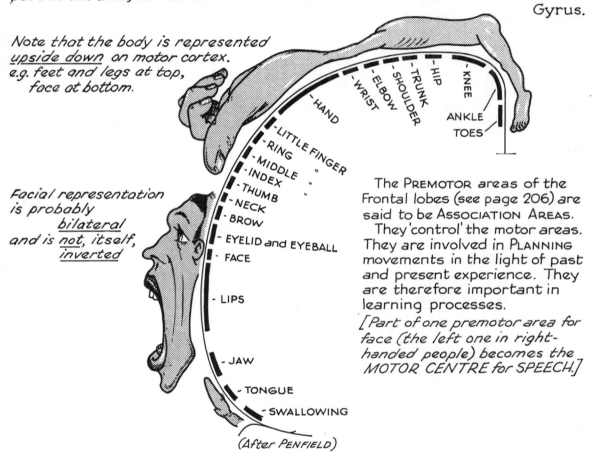

Note that the body is represented upside down on motor cortex. e.g. feet and legs at top, face at bottom.

Facial representation is probably bilateral and is not, itself, inverted

WRIST — HAND — LITTLE FINGER — RING " — MIDDLE " — INDEX " — THUMB " — NECK — BROW — EYELID and EYEBALL — FACE — LIPS — JAW — TONGUE — SWALLOWING

ELBOW — SHOULDER — TRUNK — HIP — KNEE — ANKLE — TOES

The PREMOTOR areas of the Frontal lobes (see page 206) are said to be ASSOCIATION AREAS.

They 'control' the motor areas. They are involved in PLANNING movements in the light of past and present experience. They are therefore important in learning processes.

[Part of one premotor area for face (the left one in right-handed people) becomes the MOTOR CENTRE for SPEECH.]

(After PENFIELD)

Note the large area of the Motor Cortex (and therefore the very large number of neurones) devoted to control of voluntary movements of the HANDS. This enables them to perform complicated movements and to acquire highly intricate skills: similarly with muscles involved in ARTICULATION.

251

MOTOR PATHWAYS to HEAD and NECK

Motor Nerve Cells in the Precentral Gyrus send impulses along their nerve processes in the INTERNAL CAPSULE to MIDBRAIN, PONS and MEDULLA.

At these levels fibres cross the midline. The UPPER MOTOR NEURONES synapse with cells in the NUCLEI of the CRANIAL NERVES.

The axons of these 2nd Neurones travel in the CRANIAL NERVES to the skeletal muscles of HEAD and NECK.

MIDBRAIN

PONS

MEDULLA

SPINAL CORD

OCULOMOTOR (III)

TROCHLEAR (IV)

TRIGEMINAL (V)

ABDUCENT (VI)

FACIAL (VII)

GLOSSOPHARYNGEAL (IX)

VAGUS (X) [Supplies Thoracic and Abdominal organs]

ACCESSORY (XI)

HYPOGLOSSAL (XII)

252

MOTOR PATHWAYS to TRUNK and LIMBS

The CONTROLLING CENTRES in the MOTOR CORTEX are linked by 2 Neurones with the EFFECTOR ORGANS – the VOLUNTARY MUSCLES.

The Motor Nerve Cells in the PRECENTRAL GYRUS send out messages which travel along their nerve fibres in the INTERNAL CAPSULE,

MIDBRAIN,

PONS,

MEDULLA – where most of the fibres cross to the other side and continue downwards in the LATERAL CORTICO-SPINAL TRACT.

Some fibres remain uncrossed in ANTERIOR CORTICO-SPINAL TRACT.

These UPPER MOTOR NEURONES now synapse with cells in the ANTERIOR HORN of the SPINAL CORD at various levels depending on the muscles they are destined to supply.

The AXONS of these LOWER MOTOR NEURONES travel in the SPINAL nerves to the skeletal muscles of trunk and limbs.

LATERAL CORTICO-SPINAL TRACT

ANTERIOR CORTICO-SPINAL TRACT

Note:- In this way one side of the CEREBRAL CORTEX controls activity of the muscles on the OPPOSITE side of the body.

MOTOR UNIT

The AXON of the LOWER MOTOR NEURONE divides into many branches.
Each branch ends at the MOTOR END-PLATE of a single muscle fibre.

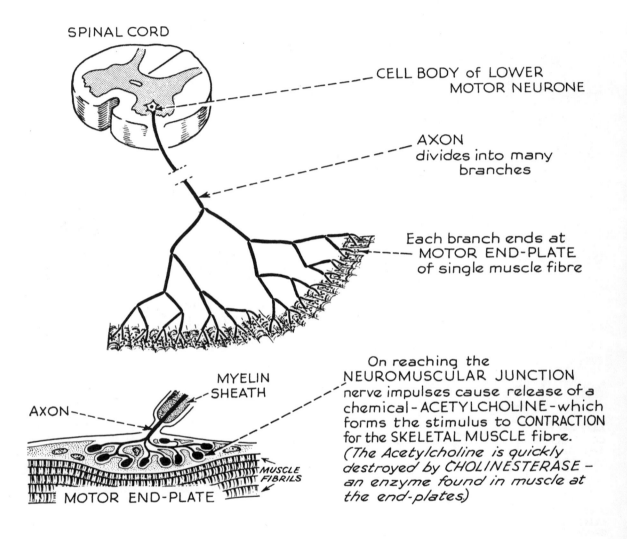

SPINAL CORD

CELL BODY of LOWER
MOTOR NEURONE

AXON
divides into many
branches

Each branch ends at
MOTOR END-PLATE
of single muscle fibre

MYELIN
SHEATH

AXON

MUSCLE
FIBRILS

MOTOR END-PLATE

On reaching the
NEUROMUSCULAR JUNCTION
nerve impulses cause release of a
chemical - ACETYLCHOLINE - which
forms the stimulus to CONTRACTION
for the SKELETAL MUSCLE fibre.
*(The Acetylcholine is quickly
destroyed by CHOLINESTERASE -
an enzyme found in muscle at
the end-plates)*

*The motor nerve with the group of muscle fibres it supplies is
known as the MOTOR UNIT. (These muscle fibres may be widely
scattered within the anatomical muscle)*

FINAL COMMON PATHWAY

Each MOTOR NEURONE in the ANTERIOR HORNS of the SPINAL CORD serves as the PATHWAY for motor impulses initiated in —
HIGHER MOTOR CENTRES in CEREBRUM *(of opposite side)*
and travelling in –

(1) *CORTICO-SPINAL TRACT*

The Motor Neurone also serves as the PATHWAY for co-ordinating corrective (restraining or facilitating) impulses discharged by—
LOWER MOTOR CENTRES in EXTRAPYRAMIDAL SYSTEM *(of same or opposite side)* and travelling in –

(2) *RUBRO-SPINAL TRACT*
from Red Nucleus *(opposite side)*

(3) *DORSAL VESTIBULO-SPINAL TRACT*
from Dorsal Vestibular Nucleus *(same side)*

(4) *OLIVO-SPINAL TRACT*
from Olivary Nucleus *(same side)*

(5) *RETICULO-SPINAL TRACT*
from Reticular Nuclei *(same side)*

(6) *VENTRAL VESTIBULO-SPINAL TRACT*
from Vestibular Nuclei *(opposite side)*

(7) *TECTO-SPINAL TRACT*
from Tectum *(opposite side)*

The Motor Neurone also receives relays of afferent impulses from other reflex centres in –
SPINAL CORD

(8) For REFLEXES of <u>same</u> segment of cord and from <u>same</u> side of cord.

(9) For REFLEXES of <u>same</u> segment but <u>other</u> side of cord and body.

(10) For REFLEXES from <u>other</u> segments of cord but <u>same</u> side of cord.

(11) For REFLEXES from <u>other</u> segments and <u>other</u> side of cord.

"FINAL COMMON PATHWAY"

Skeletal Muscle Fibres

If several sources compete for the "FINAL COMMON PATHWAY" at any one time, allied ones may reinforce each other: if incompatible, those which are initiated by PAINFUL stimuli (i.e. PROTECTIVE REFLEXES) take precedence.

EXTRAPYRAMIDAL SYSTEM

The actual performance of a VOLUNTARY MOVEMENT— initiated in the HIGHER MOTOR CENTRES in the CEREBRUM —involves co-operation by LOWER MOTOR CENTRES in the BRAIN STEM.

These Lower Centres are grouped together as the <u>EXTRAPYRAMIDAL SYSTEM</u>

<u>CORPUS STRIATUM</u>

<u>CAUDATE NUCLEUS</u>
<u>LENTIFORM NUCLEUS:</u>
PUTAMEN,
GLOBUS
PALLIDUS

These centres have important 2-way connections with each other and with HIGHER and LOWER MOTOR and SENSORY Centres

<u>CEREBELLUM</u>
(for centres and connections see pp. 258-9)

<u>MIDBRAIN and PONS</u>
RED NUCLEI
SUBSTANTIA NIGRA
RETICULAR NUCLEI
(scattered groups of neurones)
TECTUM
VESTIBULAR NUCLEI

PREMOTOR CORTEX

CAUDATE NUCLEUS

THALAMUS

LENTIFORM NUCLEUS

SUB- & HYPO-THALAMUS

RED NUCLEUS

SUBSTANTIA NIGRA

TECTUM

VESTIBULAR NUCLEUS

DENTATE NUCLEUS

RETICULAR NUCLEI

RUBRO-SPINAL TRACT

RETICULO-SPINAL TRACT

TECTO-SPINAL TRACT

VESTIBULO-SPINAL TRACT

'Suppressor areas' send impulses which influence these Lower Centres

'Receiving Centres' for 'information'

'Discharging Centres'. Impulses sent out to modify MUSCULAR CONTRACTION.

Through these widespread connections the *EXTRAPYRAMIDAL SYSTEM* exercises a STEADYING influence on MUSCLE TONE.

Concerned in Control of MUSCLE TONE and CO-ORDINATION, and in POSTURAL REFLEXES.

Some fibres carry impulses to <u>restrain</u> Lower Motor Neurone activity— others to <u>facilitate</u> it.

Interplay of impulses
↓
SMOOTH CO-ORDINATED MUSCLE MOVEMENT

Little is known about the methods for regulating this interplay of impulses converging on the MOTOR UNIT during a voluntary action.

Where part of the Extrapyramidal System is destroyed by disease varying types of RIGIDITY, TREMOR and UNCO-ORDINATED MUSCLE MOVEMENT result.

CEREBELLUM

The CEREBELLUM has 2 HEMISPHERES.
Each Hemisphere has 3 LOBES. These differ in development and function.

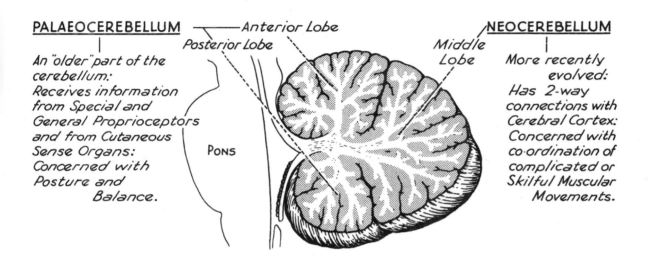

PALAEOCEREBELLUM — Anterior Lobe

Posterior Lobe

An "older" part of the cerebellum:
Receives information from Special and General Proprioceptors and from Cutaneous Sense Organs:
Concerned with Posture and Balance.

PONS

Middle Lobe

NEOCEREBELLUM

More recently evolved:
Has 2-way connections with Cerebral Cortex:
Concerned with co-ordination of complicated or Skilful Muscular Movements.

Cerebellar Cortex —
            Grey matter on surface

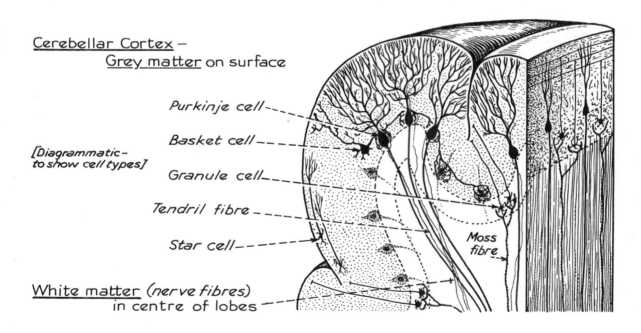

[Diagrammatic- to show cell types]

Purkinje cell
Basket cell
Granule cell
Tendril fibre
Star cell

Moss fibre

White matter (nerve fibres)
        in centre of lobes

257

CEREBELLUM

Each Cerebellar hemisphere is linked with the rest of the Nervous System through 3 bundles of nerve fibres — the Superior, Middle and Inferior Peduncles.

The Cerebellum receives information....

....from
GENERAL PROPRIOCEPTORS
*(Ventral Spino-cerebellar Tract
from Spinal Cord
- conveying impulses
from muscle-joint
proprioceptors)*

....from
CEREBRAL CORTEX
*(frontal and
temporal areas)*
- information re
"planned" movements
of muscles and
information from
SPECIAL SENSES ⟨ *Eyes*
                    Ears
*(exact pathways
unknown)*

....from
**OLIVARY
NUC.**

....from
CUTANEOUS SENSE ORGANS

SUPERIOR
PEDUNCLE

MIDLINE

NUCLEI
PONTIS

MIDDLE
PEDUNCLE

INFERIOR
PEDUNCLE

MIDLINE

GRACILE
&
CUNEATE
NUCLEI

....from
GENERAL PROPRIOCEPTORS *(Dorsal Spino-cerebellar Tract)* - information on position of limbs and movements of muscles and joints.

....from
SPECIAL PROPRIOCEPTORS
Semi-circular Canals - information re movement of head in space.
Utricle
Saccule } - information re position of head in space.

The activities of the Cerebellum are carried out beneath the level of consciousness. The information reaching it gives no sensation.

CEREBELLUM

OUTGOING PATHWAYS

In addition to the grey matter on the surface of the cerebellar cortex, there are masses of grey matter — called CEREBELLAR NUCLEI — deep within the cerebellum.

The **PALAEOCEREBELLUM** sends out messages to...
..... RED NUCLEUS of opposite side to influence the activity of the primitive Rubro-spinal motor pathway.

..... CRANIAL NERVE NUCLEI III, IV, VI and XI to co-ordinate movements and position of head with movements of eyes and trunk.

Vestibular Nucleus
Vestibulo-spinal tract to co-ordinate movements and position of head with Postural Tone of limb muscles.

The **NEOCEREBELLUM** sends out messages to....
....PRE-MOTOR region of CEREBRAL CORTEX (via THALAMUS of opposite side) to influence "Voluntary" actions.

.....RED NUCLEUS of opposite side to influence activity of the Rubro-spinal motor pathway

SUPERIOR PEDUNCLE

MIDDLE PEDUNCLE

MIDLINE

INFERIOR PEDUNCLE

NUCLEUS EMBOLIFORMIS
NUCLEUS GLOBOSUS
NUCLEUS FASTIGII

SUPERIOR PEDUNCLE

DENTATE NUCLEUS

The cerebral cortex initiates purposeful movements. During such movements proprioceptors are continually supplying information to the Cerebellum about the changing positions of muscles and joints. The Cerebellum is then responsible for collating this information and sending out impulses to bring about the smooth co-ordinated movements of different groups of muscles.

CONTROL of MUSCLE MOVEMENT

The muscle spindle is the key structure in the complex self-regulating mechanism for the control of movement of skeletal muscle.

In VOLUNTARY MOVEMENT a muscle can be made to contract by impulses reaching it by one of 2 routes:-

DIRECT PATHWAY

Impulses from HIGHER CENTRES in CEREBRAL CORTEX pass down large motor neurones of the Pyramidal tract — the alpha (α) route — to excite directly the LOWER MOTOR NEURONES → MOTOR UNITS — and lead to CONTRACTION of MUSCLE.

—This is probably the pathway involved in sudden movements

INDIRECT PATHWAY

Impulses from HIGHER CENTRES especially the CEREBELLUM pass in small motor neurones to intrafusal muscle fibres in the MUSCLE SPINDLE — the gamma (γ) route — These contract and cause stretching of spiral sensory endings which then discharge an increased number of impulses along their afferent fibres into the Spinal Cord.
This reflexly leads to the discharge of impulses in the LOWER MOTOR NEURONE and CONTRACTION of MUSCLE.

—This is probably the pathway for accurately controlled and fine movements and also for movements involved in maintenance of posture.

It is probable that the CEREBRAL CORTEX decides WHAT is to be done and the CEREBELLUM 'decides' HOW it is to be done (i.e. whether the direct or indirect pathway is to be used).

LOCOMOTOR SYSTEM

<u>BONES and MUSCLES</u> ——— concerned with <u>MOVEMENT</u> of the body.

<u>SKELETON</u>——— <u>RIGID FRAMEWORK</u> gives <u>SHAPE</u> and <u>SUPPORT</u> to body.
is <u>JOINTED</u> to permit <u>MOVEMENT</u>

FLAT BONES protect delicate organs.

LONG BONES act as levers.

SHORT BONES confer strength.

The <u>VERTEBRAL COLUMN</u> is BASIS of <u>SKELETON</u>

CLAVICLE
SCAPULA
HUMERUS
RIBS
VERTEBRAL COLUMN
ILEUM
ULNA
RADIUS
CARPALS
META-CARPALS
PHALANGES
FEMUR
PATELLA
FIBULA
TIBIA
TARSUS
METATARSALS
PHALANGES

<u>33 VERTEBRAE</u>
A series of bony rings united by intervertebral discs of cartilage

7
CERVICAL — Atlas / Axis

12
THORACIC

5
LUMBAR

5
SACRAL — Fused together

4
COCCYGEAL — Fused together

Hollow canal encloses and protects SPINAL CORD

Supports skull: permits nodding.
Allows rotation of head.
Allow bending and twisting movements of neck.

Support ribs.
Allow rotation, forward bending and some sideways movement of trunk.

Allow backward bending, sideways movement and some rotation of trunk.

Transmit weight of body to pelvic girdle and legs.

All bones give attachment to muscles.

SKELETAL MUSCLES

Bones are moved at joints by the CONTRACTION and RELAXATION of MUSCLES attached to them.

PECTORAL,
brings arm to side and across chest

BICEPS
bends elbow

FLEXORS,
bend wrist and fingers

RECTUS FEMORIS
bends hip-joint and straightens knee

ADDUCTORS of THIGH

SARTORIUS
bends knee and hip-joints and turns thigh outwards

EXTENSORS
turn foot and toes upwards

FACIAL muscles are involved in varying facial *EXPRESSION; SPEECH; MASTICATION* [*Some muscles link bone to skin.*]

The deep muscles of the THORAX, linking the ribs, contract and relax in *RESPIRATION.*

The muscles of the ABDOMEN are arranged in sheets and *PROTECT* delicate abdominal organs. They also contract to compress abdominal contents and aid in *MICTURITION, DEFAECATION, VOMITING* (and in the processes of *CHILDBIRTH* in the female).

In the LEGS are found the most powerful muscles of the body – especially those acting on the hip-joint.

Muscles which bend a limb at a joint are called **FLEXORS.**

Muscles which straighten a limb at a joint are called **EXTENSORS.**

Muscles which move a limb (or other part) away from the midline are called **ABDUCTORS.**

Muscles which move a limb (or other part) towards the midline are called **ADDUCTORS.**

SKELETAL MUSCLES

EXTENSORS
straighten wrist and fingers

TRICEPS
straightens elbow

DELTOID
raises arm

TRAPEZIUS
*raises shoulder and
pulls head back*

LATISSIMUS DORSI
*draws arm backwards
and turns it inwards
(It also draws downwards
an upstretched arm)*

The muscles of the
BACK play a large part in
maintaining erect posture

GLUTEALS
*straighten hip-joint
and move leg
outwards*

HAMSTRINGS
*bend knee and
straighten hip joint*

GASTROCNEMIUS
*bends knee and
turns foot downwards*

Some muscles
work together to
ROTATE a limb
or other part of
the body.

ACHILLES TENDON

FLEXORS
*turn foot and
toes downwards*

263

MUSCULAR MOVEMENTS

The long bones particularly form a light framework of LEVERS.
The skeletal muscles attached to them contract to operate these levers.

When a muscle contracts it shortens.

This brings its two ends closer together.

Since the two ends are attached to different bones by TENDONS one or other of the bones must move.

Two bones meet or articulate at a JOINT.

Joint surfaces are covered with a layer of smooth CARTILAGE.

To avoid friction when the two surfaces move on one another a SYNOVIAL MEMBRANE secretes a LUBRICATING FLUID.

BICEPS

TRICEPS

FLEXION

EXTENSION

The muscle which contracts to move the joint is called the PRIME MOVER or AGONIST *(the Biceps in flexion of elbow)*

To allow the movement to take place, however, other muscles near the joint must co-operate :-
The oppositely acting muscles gradually relax – these are called the ANTAGONISTS and exercise a "braking" control on the movement. *(e.g. the EXTENSORS – chiefly Triceps - in flexion of the elbow.)*
Other muscles steady the bone giving "origin" to the Prime Mover so that only the "insertion" will move— these muscles are called FIXATORS.
Still other muscles help to steady, for most efficient movement, the joint being moved — called SYNERGISTS.

When the elbow is straightened the reverse occurs:-
TRICEPS, the Prime Mover, CONTRACTS;   BICEPS, the Antagonist, RELAXES.

RECIPROCAL INNERVATION

The co-ordinated group action of muscles is made possible by the many synaptic connections between CONNECTOR NEURONES of the INGOING or PROPRIOCEPTIVE NEURONES of one muscle group and the OUTGOING or MOTOR NEURONES of the functionally opposite group of muscles.

This is shown diagrammatically for the alternating contractions and relaxations of the Flexors and Extensors of one knee in Walking :-

REFLEX FLEXION MOVEMENT

FLEXOR muscles <u>contract</u>   EXTENSOR muscles <u>relax</u>

As limb flexes ⟶ *Muscle Spindles of Extensors are stretched*

REFLEX EXTENSION MOVEMENT

REFLEX <u>relaxation</u> of FLEXORS   REFLEX <u>contraction</u> of EXTENSORS

Muscle Spindles of Flexors are stretched ⟵ As limb extends

REFLEX FLEXION MOVEMENT

REFLEX <u>contraction</u> of FLEXORS   REFLEX <u>inhibition</u> again of EXTENSORS

The rhythmic alternating reflex movements of walking

i.e. The FINAL COMMON PATHWAY *is commandeered alternately by "exciting" or "inhibiting" messages.*

RECIPROCAL INNERVATION is probably due to division (within the SPINAL CORD) of the CONNECTOR NEURONES for each AFFERENT NERVE — e.g. one branch excites (+) extensor motor neurone; the other branch inhibits (−) the flexor motor neurone.

Thus :-
When limb is flexed-

	AFFERENT	CONNECTOR	FINAL COMMON PATH: MOTOR NEURONE.
Muscle Spindles in Extensors stretched	*discharges* ⟶	*facilitates contraction* *inhibits activity* +	*of Extensor muscle* ⟶ +
Muscle Spindles in Flexors relaxed	⟶	*facilitates activity* *facilitates relaxation* −	*of Flexor muscle* −

RECIPROCAL INNERVATION helps make possible self-regulation of the rhythmical movements.

SKELETAL MUSCLE
and the MECHANISM of CONTRACTION

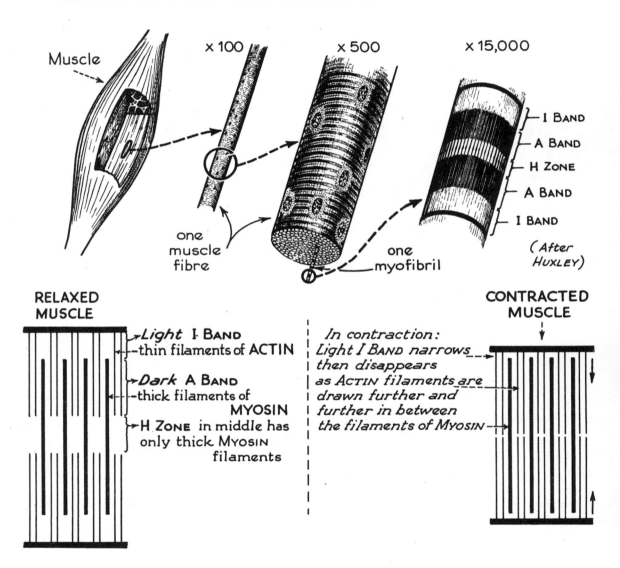

Muscle

×100

×500

×15,000

one muscle fibre

one myofibril

I BAND
A BAND
H ZONE
A BAND
I BAND

(After HUXLEY)

RELAXED MUSCLE

CONTRACTED MUSCLE

Light I BAND
—thin filaments of ACTIN

Dark A BAND
—thick filaments of MYOSIN

H ZONE in middle has only thick MYOSIN filaments

In contraction:
Light I BAND narrows
then disappears
as ACTIN filaments are
drawn further and
further in between
the filaments of MYOSIN

Energy for contraction is derived from glucose and fat in the mitochondrion (*page 6*). The energy is transported from the mitochondrion to the contractile filaments in ATP. ATP splits readily into ADP and phosphate, releasing its trapped energy where needed (*page 42*).

266

AUTONOMIC
NERVOUS SYSTEM
and
CHEMICAL TRANSMISSION
at NERVE ENDINGS

AUTONOMIC NERVOUS SYSTEM

The Autonomic Nervous System is concerned with maintaining a <u>Stable Internal Environment</u>. It governs functions which are normally carried out below the level of consciousness.

It has 2 separate parts – <u>PARASYMPATHETIC</u> and <u>SYMPATHETIC</u>. Both have their Highest Centres in the Brain.

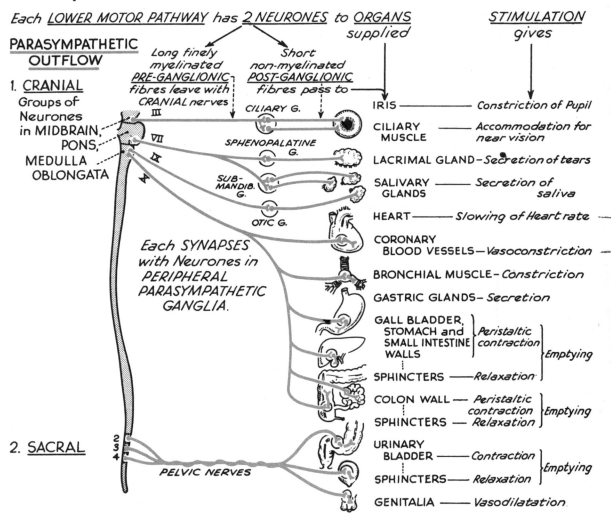

Each <u>*LOWER MOTOR PATHWAY*</u> has <u>*2 NEURONES*</u> to <u>*ORGANS*</u> supplied

STIMULATION gives

PARASYMPATHETIC OUTFLOW

Long finely myelinated <u>*PRE-GANGLIONIC*</u> fibres leave with *CRANIAL* nerves

Short non-myelinated <u>*POST-GANGLIONIC*</u> fibres pass to

1. <u>CRANIAL</u>
Groups of Neurones in MIDBRAIN, PONS, MEDULLA OBLONGATA

III
VII
IX
X

CILIARY G.
SPHENOPALATINE G.
SUB-MANDIB. G.
OTIC G.

IRIS —————— *Constriction of Pupil*

CILIARY MUSCLE ——— *Accommodation for near vision*

LACRIMAL GLAND – *Secretion of tears*

SALIVARY GLANDS ——— *Secretion of saliva*

HEART ———— *Slowing of Heart rate*

CORONARY BLOOD VESSELS – *Vasoconstriction*

BRONCHIAL MUSCLE – *Constriction*

GASTRIC GLANDS – *Secretion*

Each SYNAPSES with Neurones in PERIPHERAL PARASYMPATHETIC GANGLIA.

GALL BLADDER, STOMACH and SMALL INTESTINE WALLS } *Peristaltic contraction* } *Emptying*

SPHINCTERS ——— *Relaxation*

COLON WALL — *Peristaltic contraction* } *Emptying*

SPHINCTERS ——— *Relaxation*

2. <u>SACRAL</u>

2
3
4

PELVIC NERVES

URINARY BLADDER ———— *Contraction* } *Emptying*

SPHINCTERS——— *Relaxation*

GENITALIA ——— *Vasodilatation.*

General stimulation of the Parasympathetic promotes vegetative functions of the body. Stimulation of the Parasympathetic and inhibition of Sympathetic have the same overall effects.

AUTONOMIC NERVOUS SYSTEM

The Parasympathetic and Sympathetic systems normally act in balanced reciprocal fashion.   The activity of an organ at any one time is the result of the two opposing influences.

Each <u>LOWER MOTOR PATHWAY</u> has <u>2 NEURONES</u> to <u>ORGANS</u>   <u>STIMULATION</u>
supplied   gives

SYMPATHETIC
OUTFLOW

Short finely myelinated <u>PRE-GANGLIONIC</u> fibres leave with MOTOR ROOTS of SPINAL NERVES. These <u>SYNAPSE</u> with Neurones in the <u>PARAVERTEBRAL GANGLIA</u> or the <u>PREVERTEBRAL GANGLIA</u> or directly with Cells in <u>SUPRARENAL MEDULLAE</u>.

Long POST-GANGLIONIC fibres pass to

PARAVERTEBRAL GANGLIA

T1

From Neurones in LATERAL HORNS of THORACIC and LUMBAR parts of SPINAL CORD

COELIAC G.

SUP. MESENTERIC G.

L2

INF. MESENTERIC GANGLION

PREVERTEBRAL GANGLIA

SWEAT GLANDS ———— Secretion
SMOOTH MUSCLE in Skin — Contraction
BLOOD VESSELS in Skin —— Constriction

IRIS ———— Dilatation of Pupil
BLOOD VESSELS of Head – Vasoconstriction
BLOOD VESSELS of Skeletal Muscle in upper limbs – Relaxation
HEART ——— Heart Rate quickens
CORONARY VESSELS – Vasodilatation
BRONCHIAL MUSCLE ——— Relaxation
STOMACH WALL ——— Relaxation
SPHINCTERS ——— Constriction
BLOOD VESSELS of Abdomen —— Vasoconstriction
LIVER – Mobilization of Liver GLYCOGEN
SUPRARENAL MEDULLAE – Secretion of ADRENALINE
SMALL INTESTINE and COLON WALLS ——— Relaxation
SPHINCTERS ——— Constriction
URINARY BLADDER —— Relaxation WALL
SPHINCTER ——— Constriction
GENITALIA ——— Vasoconstriction

General stimulation of Sympathetic results in mobilization of resources to prepare body to meet emergencies.   Stimulation of the Sympathetic and inhibition of the Parasympathetic have the same overall effects.

AUTONOMIC REFLEX

AUTONOMIC CENTRES in the BRAIN and SPINAL CORD receive SENSORY INFLOWS from the VISCERA. *(less is known about their exact pathways than about MOTOR OUTFLOWS.)* Some of the SENSORY NEURONES convey information about events in the viscera to HIGHER AUTONOMIC CENTRES which send impulses to modify the activity of ——————

Both VISCERAL and SOMATIC AFFERENTS serve as AFFERENT PATHWAYS for AUTONOMIC REFLEXES by means of which much of the nervous regulation of vegetative functions is carried out below the level of consciousness.

e.g. *The Simplest Autonomic Reflex arc:-*

AFFERENT PATHWAY

Like Somatic Sensory Neurone the cell body lies in the Posterior Root Ganglion of Spinal Nerve (or in Sensory Ganglion of a Cranial Nerve)

CEREBRAL CORTEX
THALAMUS
HYPOTHALAMUS
MEDULLA

LOWER AUTONOMIC CENTRES in BRAIN STEM and SPINAL CORD

EFFERENT PATHWAYS *involve chain of 2 neurones*

PRE-GANGLIONIC and POST-GANGLIONIC

LATERAL GANGLION of SYMPATHETIC CHAIN

COLLATERAL GANGLION in ABDOMEN

Examples of RECEPTORS

Sensory Nerve Endings in
SKIN
WALLS of BLOOD VESSELS
VISCERAL MUSCLE
etc.

Smooth Muscle of VISCERA and Blood Vessels of VISCERA

Smooth Muscle of BLOOD VESSELS

SWEAT GLANDS

ARRECTORES PILORUM
etc.

EFFECTORS

The Autonomic Reflex Arc differs from the Somatic Reflex Arc mainly in that it has 2 Efferent neurones. Transmission of impulse from Afferent to Efferent probably involves one or more connecting neurones.

CHEMICAL TRANSMISSION at NERVE ENDINGS

When an impulse passes along a nerve, sodium enters the fibre and potassium leaves it. When this process reaches the synapse between neurone and neurone or between neurone and effector organ a chemical substance is liberated. This bridges the gap and forms the stimulus to alter the membrane potential and initiate in the next neurone (or in the effector) the wave of depolarization and altered membrane permeability with ion exchange which becomes the transmitted nerve impulse.

Some nerves when stimulated liberate an acetylcholine-like substance. These are called CHOLINERGIC. Others liberate nor-adrenaline. These are called ADRENERGIC.

CENTRAL NERVOUS SYSTEM

AUTONOMIC NERVOUS SYSTEM

Chemical transmission in sensory neurones is still open to speculation. The chemical transmitter (if any) is not known. In certain sites it may be Acetylcholine

[It has been suggested (Feldberg & Vogt) that alternate neurones in the C.N.S. are Cholinergic and non-Cholinergic]

SENSORY CORTEX   MOTOR CORTEX

non-cholinergic(?)
3RD Sensory Neurone from Thalamus

cholinergic (?)
2ND Sensory Neurone from Cuneate and Gracile Nuclei

Upper Motor Neurone (Pyramidal fibres) *non-cholinergic(?)*

1ST Sensory Neurone *non-cholinergic(?)*

Lower Motor Neurone is *cholinergic*

RECEPTORS

Motor End-plates

Skeletal Muscle

PARASYMPATHETIC   SYMPATHETIC

Preganglionic fibre *cholinergic*

cholinergic
Adrenal Medulla

Preganglionic fibre *cholinergic*

Postganglionic fibre *cholinergic*

Postganglionic fibre *adrenergic*

Cardiac M.

Smooth M.

-- EFFECTORS --

Glands

There are some exceptions to the above general rule. Some Sympathetic and Parasympathetic nerves appear to include both cholinergic and adrenergic fibres. e.g. POSTGANG. SYMP. fibres to SWEAT GLANDS are CHOLINERGIC

There are thought to be other chemical substances secreted by neurones which may act as SYNAPTIC TRANSMITTERS to regulate certain aspects of cerebral activity. Amongst these are SEROTONIN and other SYMPATHOMIMETIC AMINES.

INDEX

Mucin—*continued*
    stomach, 58
    trachea, 118
Mucosa
  in digestive tract, 67
    large intestine, 72
    oesophagus, 56
    small intestine, 66
    stomach, 58
    muscularis mucosae, 67
Mucous glands, 9
  in trachea, 118
  oesophagus, 56
  salivary glands, 54
Multipolar neurones, (15), 16
Muscles
  action of insulin, 173
  chemical transmission in, 271
  contraction of, 264, 266
  development, 19
  effect of adrenaline on, 161
  eye, 224-226
  heat production in, 43, **44, 45**
  histology, 12, 14, 266
  in diet, 34, 35
  -joint sense, 220, 244
    pathways, 245, 258
  of mastication, nerve supply to, 210
  reciprocal innervation of, 265
  release of energy in, 42, 266
  respiratory, 120, 121
  innervation, 130
  skeletal, 262, 263, 266
  spindle, 244
    in reciprocal innervation, 265
    stretch reflexes, 215
    role in control of muscle movement, 260
  tone, 44, 45
    role of extrapyramidal system, 256
    role of proprioceptors, 244
  utilization of glucose by, 40
Muscular arteries, 91
  blood pressure in, 92
  nervous regulation of, 95
Muscularis mucosae, 67
  in oesophagus, 56
  'villus pump', 70
Muscular veins, 91
  blood pressure in, 92
  'tone' in, 99
Myelencephalon, 205
Myelin, 15, 16
  and conduction of nerve impulse, **213,** 254
Myeloblast, 107
Myocardium, 80
Myoepithelial cells, mammary glands, 169
Myofibril, 266
Myogenic movement, 69, 73, 75
Myometrium, **181,** 186, 187, 190, 192, 194-196
Myopia, 229
Myosin, 266
Myxoedema, 152

Nasal passages, 117, 221

Naso-lacrimal duct, 221, 225
Negative intra-thoracic pressure, 120, 121
Neocerebellum, 257, **259**
Nephrons, 135
  clearance of diodone, 140
    inulin, 138
    urea, 139
Nerves
  adrenergic, 271
  cholinergic, 271
  cranial, 210
  histology, 15-17
  nature of impulse, 212, **213**
  spinal, 211
Nervous tissues, **15, 16, 17,** 20, **204**
  development of, 19, 46, 205
Neurilemma, 15, 16
Neurogenic movement, 69, 73, 75
Neuroglia, 15, 16
Neurohormones
  antidiuretic hormone, 168, 170, 171
  oxytocin, 168
Neuromuscular excitability
  role of blood calcium and parathyroid, 154, 155
Neuromuscular junction, 254
  chemical transmission at, 271
Neuromuscular transmission, 254, 271
Neurones, **15,** 16, 17
  arrangement of, 219
  nature of nerve impulse, 213
  synapse, 212
Neurosecretion
  hypothalamus, 168, 170, 171, 172
Neutral (unsplit) fat, (triglycerides), 27
  absorption of, 41, 70, 71
  digestion of, 66
    role of bile, 64
Neutrophil, 11, 107
Newborn
  circulation, 193
  energy requirements of, 49
  haemolytic disease of, 110
  haemopoiesis in, 106
  thymus in, 173
  urine in, 141
Nicotinamide nucleotides, 42
Nicotinic acid, 37
Nitrogen, 25, 28, 29
  cycle, 32
  retention, role of growth hormone, 163-165
Nodal tissue (heart), **85,** 86
Node of Ranvier, 15, 16
  in conduction of nerve impulse, 213
Non-granular leucocytes, 11
  formation of, 107, (112), (113)
Noradrenaline, 160
  in chemical transmission, 271
  in gut, 74
Normoblast, 107
Nose, 117, 210, 220, **221**
Nucleases, 62, 66
Nucleic acids, 29
  digestion of, 62, 66
  in leucopoiesis, 106

Nucleolus, 6, 7
Nucleoplasm, 3, 6
Nucleoproteins, 29
Nucleosidases, 66
Nucleosides, 66
Nucleotidases, 66
Nucleotides, 29, 66
Nucleus
  caudate, 207, 256
  cochlear, 240
  cranial nerve nuclei, 245, 247, 252
  cuneate, 245, 249
  dentate, 259
  division of, 7, 18
  Edinger-Westphal, 228, 236
  emboliformis, 259
  fastigii, 259
  globosus, 259
  gracile, 245, 249
  histology, 6, 7
  lentiform, 207, 256
  nucleic acids, 29
  olivary, 240, 258
  red, 256, 259
  solitarius, 223
  supra-optic, 144, 145, 168-171
  vestibular, 243, 256
Nutrition, 4, 8, 22, **33,** 34-37
  factors in haemopoiesis, 106
  influence on growth, 46
Nystagmus, 242

Obesity
  in Cushing's syndrome, 159, 166
  in Fröhlich's syndrome, 164
Obligatory reabsorption, 171
Occipital lobes, 206-208
  centres for eye reflexes, 227
  centres for vision, 234, 235
Oculomotor nerve (III), 210, (226), (227), 252
  parasympathetic outflow to ciliary muscle, (228), 268
Oddi, sphincter of, 62
Oedema
  in overactivity of suprarenal cortex, 159
Oesophagus, 52, 53, **56,** (154)
  innervation, 75
  lining, 10, 56
  swallowing, 57
Oestrogen, 180, **182, 185,** 186, **187,** 189-192, 196, 199, 200, 201
  action on mammary glands, 196-198
  after childbirth, 195
  at menopause, 196, 200
  at puberty, 184
  effect on anterior pituitary, 163
  in male, 157, 179
  suprarenal cortex, 157-159
Olfactory
  bulb, (210), 221
    development of, 205
  centres, 221
  epithelium, 221
  nerve, 210, 221
  receptors, 221